EYEWITNESS
PLANETS

A young rocky planet
bombarded by spacerocks

A meteorite fragment in false colour

The Moon

EYEWITNESS
PLANETS

Written by
Carole Stott

Pluto, a dwarf planet

Saturn V rocket
blasts off the
launchpad

Earth's Pacific Ocean

DK | Penguin
Random
House

The author dedicates
this book to Elizabeth Rosa Lee

Consultant Dr Jacqueline Mitton

DK LONDON
Senior editors Camilla Hallinan, Jenny Sich
Senior designer Spencer Holbrook
Jacket editor Claire Gell
Jacket design development manager Sophia MTT
Producer, pre-production David Almond
Producer Gary Batchelor
Managing editor Francesca Baines
Managing art editor Philip Letsu
Publisher Andrew Macintyre
Associate publishing director Liz Wheeler
Art director Karen Self
Design director Philip Ormerod
Publishing director Jonathan Metcalf

DK DELHI
Senior art editor Sudakshina Basu
Editor Priyanka Kharbanda
Jacket designer Garima Sharma
Senior DTP designer Harish Aggarwal
DTP designers Rajesh Singh Adhikari,
Syed Md Farhan, Pawan Kumar
Picture researcher Sumedha Chopra
Picture research assistant Esha Banerjee
Managing jackets editor Saloni Singh
Picture research manager Taiyaba Khatoon
Pre-production manager Balwant Singh
Production manager Pankaj Sharma
Managing editor Kingshuk Ghoshal
Managing art editor Govind Mittal

First published in Great Britain in 2017
by Dorling Kindersley Limited
80 Strand, London WC2R 0RL

A WORLD OF IDEAS:
SEE ALL THERE IS TO KNOW

www.dk.com

Earth-orbiting satellite

Close-up, false-colour view
of Saturn's north pole

Jupiter outweighs all seven other planets combined

Contents

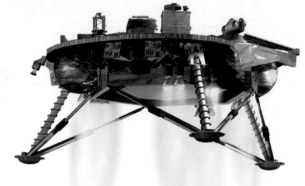

Phoenix lander

Planet Earth and its neighbours

Part of the family of space objects called the Solar System, Earth and seven more planets formed about 4.6 billion years ago. They each orbit the Sun, as do billions of smaller bodies such as asteroids, and others far beyond the planets.

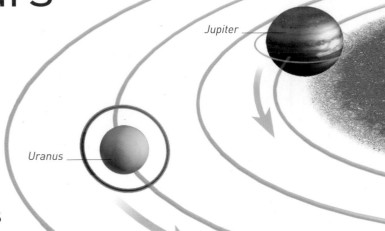

Jupiter

Uranus

Mercury

Planetary Solar System
The Sun is our local star. Its gravity holds the Solar System together. Following elliptical orbits – stretched, circular paths – around the Sun, the eight planets and the asteroids together orbit in a disc-like plane that extends about 4.5 billion km (2.8 billion miles) out from the Sun.

Inner planets
The four planets closest to the Sun are also the Solar System's smallest. Often called the inner or rocky planets, all four are balls of rock and metal, but with very different surfaces. Earth has life and oceans, Mars is a frozen desert, Venus has a volcanic surface, and Mercury is covered with craters.

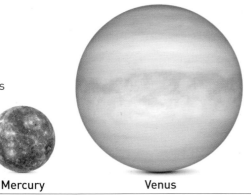

Mercury   Venus   Earth   Mars

Inner planets

Outer giants
Known as the giants, Jupiter, Saturn, Uranus, and Neptune are the outermost and largest of the Solar System planets. They are also the most massive, being made of the most material – much of it hydrogen. Each has a thick atmosphere, rather than a solid surface, and a ring system.

Jupiter   Saturn   Uranus   Neptune

Outer giants

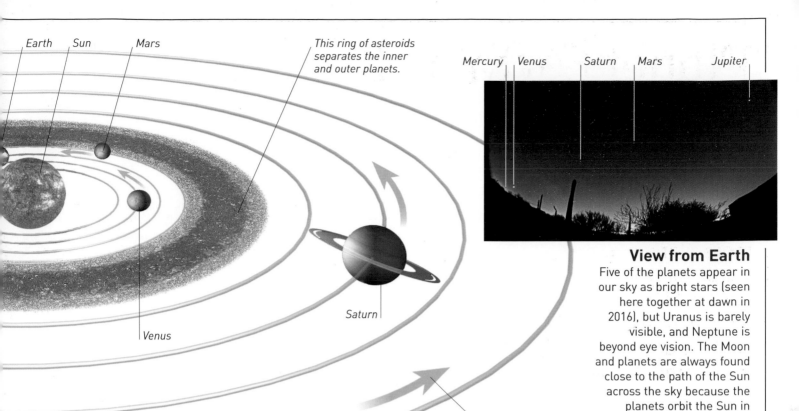

Earth Sun Mars

This ring of asteroids separates the inner and outer planets.

Venus

Saturn

Anticlockwise direction of orbit

Neptune

View from Earth

Mercury Venus Saturn Mars Jupiter

Five of the planets appear in our sky as bright stars (seen here together at dawn in 2016), but Uranus is barely visible, and Neptune is beyond eye vision. The Moon and planets are always found close to the path of the Sun across the sky because the planets orbit the Sun in the same plane.

A rocky planet forms in the inner, hotter part of the disc.

Birth of the Solar System

Titan, Saturn's largest moon

The Solar System formed from a spinning cloud of gas, dust, and ice particles. As gravity pulled material inwards, the central region became denser and hotter, forming our star about 5 billion years ago. Unused material settled into a disc around the Sun (right) and formed the planets.

A giant planet takes shape in the outer region.

The young Sun starts to produce energy in its core.

Earth's axis tilts by 23.4°.

23.4°

Plane of Earth's orbit

Earth's spin is eastwards (anticlockwise seen from above the North Pole).

Planetary moons

There are more than 170 moons in the Solar System. Earth has one, and Mars has two; Mercury and Venus have none. All four giants have large families of moons – Jupiter and Saturn have more than 60 each, including Saturn's Tethys and Titan (above).

Orbit and spin

Planets closest to the Sun orbit the quickest – the further from the Sun, the longer an orbit takes. As each planet travels along its orbit, it spins on an axis that is tilted at an angle to the plane of orbit. No planet spins precisely upright (90 degrees to its plane of orbit). Earth's axis tilts by 23.4 degrees.

Structure

Beyond Neptune is the Kuiper Belt of rock and ice bodies, including the dwarf planets Pluto and Eris. More distant still is Sedna, at more than 900 times the Earth-Sun distance. Surrounding all of this is the Oort Cloud, home of comets, which marks the limit of the Sun's influence.

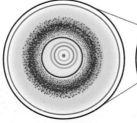

From the Sun (centre) to Jupiter (orange orbit)

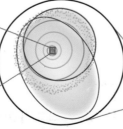

From Saturn to Pluto (purple) and Eris (red)

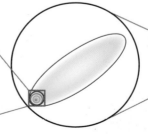

Beyond the planets, Sedna's orbit (red)

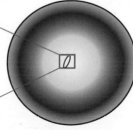

Oort Cloud surrounds all other objects

What is a planet?

For centuries, people have labelled Earth and other huge, round bodies orbiting the Sun as planets. Recent discoveries have led to a formal definition, and two new classes – dwarf planets and exoplanets. Just as Earth and seven more planets orbit the Sun, exoplanets orbit other stars. There may be tens of billions of exoplanets orbiting stars in the Milky Way Galaxy.

The planet Mercury – false colour highlights different rock types

Defining a planet

The word "planet", from the Ancient Greek for "wanderer", was used to describe so-called wandering stars – planets moving across the starry sky. In 2006, the International Astronomical Union (IAU) defined a planet as a body orbiting the Sun that is massive enough for its gravity to make it nearly round, and which has cleared other objects out of its path.

Dwarf planet

Discovered in 1930, Pluto was classed as a planet. But then Eris, a body more distant and seemingly larger than Pluto, was found in 2005. Both are far smaller than the planets and follow elongated orbits among other objects. In 2006, the IAU introduced the class of dwarf planet. A dwarf planet orbits the Sun and is massive enough for its gravity to make it round, but has not cleared the neighbourhood around its orbit.

Pluto with its moon Charon (upper left)

The habitable zone

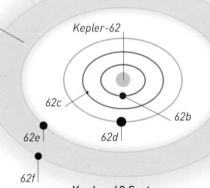

Kepler-62

62c
62e
62f
62d
62b

Kepler-62 System

The habitable zone

Earth is in the Sun's habitable zone, a region where liquid water can exist and allow life to form and flourish. Any closer to the Sun, Earth would be so hot that its water would evaporate away. Any further, Earth would be so cold, its water would freeze. Exoplanets in habitable zones could harbour life. Two planets, 62e and 62f, orbit the star Kepler-62 within its habitable zone.

The orange Sun-like star HD 189733 has one known exoplanet.

Planet HD 189733b – a gas giant larger than Jupiter – orbits its star in 2.2 days.

Exoplanet

Exoplanets are small compared to their stars, which also outshine them – making them doubly difficult to detect. We know of more than 3,500 exoplanets, and these orbit stars relatively close to the Sun. In August 2016, astronomers confirmed that our closest star – Proxima Centauri, around a quarter of a million times further from the Sun than Earth is – has an Earth-sized planet.

Naming the planets

The planets were named in ancient times after Greek or Roman gods, and we still use their Greek or Roman names today. For instance, Venus is named after the Roman goddess of love (left). Of the two planets discovered in modern times, Uranus is named after the Greek god of the sky, and Neptune after the Roman god of the sea.

American astronomer Carl Sagan pioneered the search for life on other planets

The Solar System is in an arm of the Milky Way Galaxy.

The world of astronomy

Planetary astronomers are those who focus on the Solar System's planets. They are interested in the planets' origin and evolution as well as where and what they are. By sharing their findings, they form our present understanding of these worlds.

Our place in the Universe

The Sun and its planets lie within the Milky Way Galaxy, a vast system of stars, gas, and dust that has at least 200 billion stars. More than 2,600 of those close to us have exoplanets. Astronomers think at least one in every 20 stars has exoplanets orbiting it. And there could be as many stars with exoplanets in the Universe's other 200 billion or more galaxies.

Changing worlds

Early on, the planets were bombarded by spacerocks. These fell through the atmospheres of the giant planets, but scarred the rocky inner planets. Over millions of years, volcanism, land movement, water, wind, and atmospheric changes continued to shape their landscapes.

Rocky balls

The rocky planets took shape as material left over from the Sun's formation clumped together into four hot, molten balls. As they cooled, they settled into layers, creating a rocky surface and metal core. Spacerocks crashed into the young planets, forming craters. Inside the planets, heat from radioactivity made the rock molten and some of it flowed out in volcanic eruptions.

Cratering

Millions of impact craters scar the rocky planets and their moons. They form when spacerocks smash into the surface and gouge out material. The largest craters are hundreds of kilometres across. The rate of impact has slowed since the most intense period of bombardment, 3.5 billion years ago.

Mars's Santa Maria Crater is about the size of a football pitch

Spacerocks crash into a newly formed planet, piercing its thin crust and releasing molten rock.

Volcanism

When young, the rocky planets and some moons were affected by volcanism. Molten rock erupted and flowed across landscapes, filling craters. Over time, the bodies cooled and volcanism slowed. Earth is still volcanic, but its crust is thickening and volcanism will eventually stop.

Tungurahua volcano, Ecuador

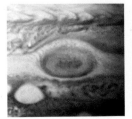

1995   2009   2014

Giant planets

As the young giant planets shrank, their cores grew hotter and their upper atmospheres formed bands where storm clouds rage. In the last few years, Jupiter's long-lived Great Red Spot (above) has been shrinking by at least 200 km (124 miles) each year.

Hot Venus, cool Earth

Young Venus and Earth both had a thick atmosphere of carbon dioxide and water. Closer to the Sun, Venus's water was broken up by the Sun's energy, leaving a thick carbon-dioxide atmosphere (above). On cooler Earth, the atmosphere thinned as water rained on to the surface and carbon dioxide gas combined into rocks that locked it in. Later, plants released oxygen into the atmosphere.

Surface building

The surfaces of the rocky planets are shaped by large-scale movements in their crusts over millions of years. As the young planets cooled and shrank, their surfaces cracked. Mercury's cliff Carnegie Rupes formed in this way. A false-colour image shows it at the edge of the higher ground (red) cutting across cratered terrain (blue).

Low land in Mars's northern hemisphere may have held an ocean of water.

Erosion

Landscapes change through erosion – the wearing away of surface material by water and wind. Pounding ocean waves change shorelines, rivers carve valleys, and glaciers pull underlying rock with them. Wind picks up material and deposits it at new sites. It also wears away rock surfaces, slowly sculpting new shapes (left).

The sandstone rock of Delicate Arch, Utah, USA, has worn away, forming a 19.5-m- (64-ft-) high hole.

Liquid water

As young Earth cooled, atmospheric steam condensed, water fell, and oceans formed. Young Mars also had liquid water in lakes and seas, but the planet has since cooled. Some water escaped via its upper atmosphere and the rest became frozen.

Skywatching

Humans have always watched the sky. The first people to study it made patterns from the stars and identified five planets. They observed the Sun, Moon, and planets moving against the background sky. The introduction of the telescope four centuries ago revealed many more space objects. Today, we know that the Solar System is a tiny part of one galaxy in a vast Universe of galaxies.

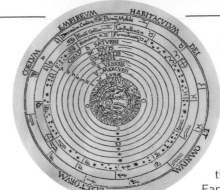

Heart of the Universe
Earth was long thought to be at the centre of the Universe, orbited by the Sun, Moon, and planets (left). In 1543, Polish scholar Nicolaus Copernicus realized that the Sun is at the centre and that the planets, including Earth, all orbit the Sun.

Using the telescope
Early telescopes – such as Italian astronomer Galileo Galilei's (replica above) – had lenses and were as powerful as a pair of simple binoculars today. From 1609, Galileo saw mountains on the Moon, four moons of Jupiter, and Venus's phases. Today's telescopes use mirrors to collect the light from distant objects. The bigger the mirror, the more we see.

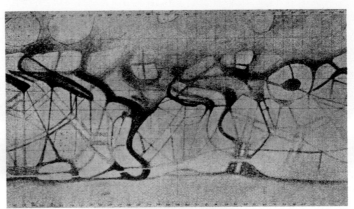

Surface features
As telescopes improved, they revealed details on planets' surfaces. In the 1870s, Italy's Giovanni Schiaparelli drew maps (left) of Mars's dark, linear features, which he called *canali* (channels). Others then mistook them for canals built by a Martian civilization – in fact, they are an optical illusion.

First observations
Ancient peoples watched the Sun rise and set each day, and the Moon change shape from night to night. They used these daily changes and the yearly movement of the Sun to keep track of time. Around 4,000 years ago, the Babylonians drew the first constellations – imaginary patterns around stars – and recorded the movements of the planets. At about the same time, Ancient Britons started building Stonehenge (below). Its stones align with the rising or setting Sun at certain times of year.

Sizing up the Solar System

In the early 17th century, German mathematician Johannes Kepler showed that planets orbit in an ellipse (an elongated circle), and those near the Sun orbit faster than those further away. A hundred years later, using English scientist Isaac Newton's laws of gravity, astronomers worked out the masses of the planets – the amount of material they are made of. Jupiter's mass is almost 2.5 times that of the other seven combined.

Expanding Universe

Until the 18th century, the Solar System was thought to end at Saturn. This changed with the discoveries of Uranus, Neptune, and Pluto. We now know of many planetary systems in the Milky Way Galaxy, itself just one of many galaxies (left).

Astrophotography

Astronomers developed the telescope so that they could see further and in more detail. When photography was invented in the mid-19th century, they turned cameras to the Moon (shown here in 1851), and then later to comets, stars, and planets. In the 1980s, astronomers began to use the CCD (charge-coupled device), the electronic chip used in the first digital cameras. Today, a large CCD camera is still a regular feature on telescopes and spacecraft.

Observatories

Several telescopes – each protected within its own building – together make an observatory. At their mountain-top locations, the air is dark, still, dry, and thin, giving the clearest possible view into the Universe. The laser beam shown above shoots out of a telescope at the Paranal Observatory in the Atacama Desert, Chile. The beam is used to correct any image blurring caused by turbulence in the Earth's atmosphere.

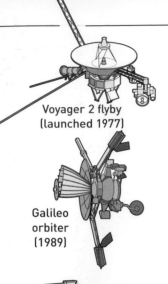

Voyager 2 flyby
(launched 1977)

Galileo
orbiter
(1989)

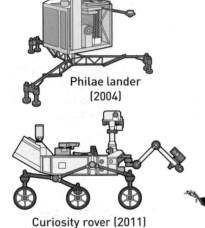

Philae lander
(2004)

Curiosity rover (2011)

Space age exploration

Robotic spacecraft have been exploring the Solar System since 1959. Far from home, in conditions no human could endure, they have investigated the planets, a host of moons, two dwarf planets, asteroids, comets, and the Sun. About the size of a family car, they carry scientific instruments that test conditions on other worlds, and transmit their findings home, making far distant worlds familiar.

Four types of robotic explorer

Flyby craft use their instruments as they travel past a planet or moon. Orbiters circle around their target – some release probes into the planet or moon's atmosphere. Landers touch down, staying where they landed, and rovers drive around visiting a mix of locations.

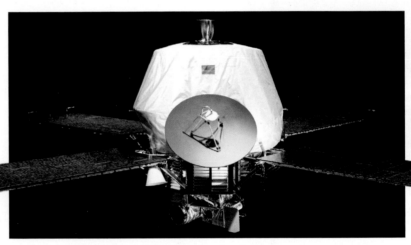

Mariner missions
Between 1962 and 1973, US Mariner missions made the first flybys of Venus, Mars, and Mercury. Mariner 9 (above) was the first craft to orbit another planet, arriving at Mars in 1971. The final mission, Mariner 10, was the first to visit two planets, Venus and Mercury.

Early exploration
The first missions to another world were the Luna craft sent by the Soviet Union to the Moon. Luna 1 was the first to leave Earth's gravity, in 1959. Luna 9 was the first to soft land on the Moon, in 1966. Lunokhod 1 (below) was the first rover to explore the Moon. It landed in 1970 and roved across 10.5 km (6.5 miles) of its surface.

Cameras provided views of Mars's north polar region.

A meteorological station made daily weather recordings on Mars.

Artist's impression of Phoenix landing in 2008

Landing craft
Spacecraft use parachutes and small rockets to control their descent and make a soft (controlled) landing. The first soft landing on a planet was made by Venera 7 on Venus in 1970, but it survived for just under an hour in the corrosive atmosphere. Mars is more hospitable – four craft have successfully landed and worked there for longer periods.

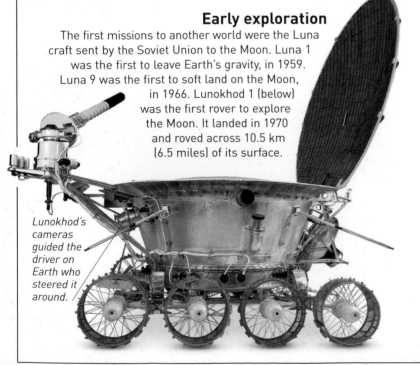

Lunokhod's cameras guided the driver on Earth who steered it around.

Sending astronauts to the Moon

Twelve American men have walked on the surface of the Moon, arriving two at a time on Apollo landing craft. The first, Apollo 11's Eagle, touched down on 20 July 1969. Just over six hours later, on 21 July, Neil Armstrong became the first man to step onto the lunar surface, followed by Buzz Aldrin. Their trip to the Moon and back was 1.5 million km (953,054 miles).

Buzz Aldrin on the Moon, photographed by Neil Armstrong

The Apollo 11 spacecraft was carried inside the upper part of the rocket.

Lower sections contained fuel and engines. These detached and fell away as the rocket climbed higher and reached Earth orbit.

A Saturn V rocket blasted off from Cape Canaveral, USA, on 16 July 1969, launching Apollo 11 on its journey to the Moon.

Bristling with equipment

Each spacecraft carries a dozen or so scientific instruments. These usually include several cameras, as seen above on the Curiosity rover on Mars. In this selfie taken by another camera, the large round eye is ChemCam, which includes a laser and telescopic camera. Below are two rectangular-shaped cameras, and at either side of them, a pair of navigation cameras.

A 34-m (112-ft) antenna near Canberra, Australia, sends commands to – and receives data from – spacecraft.

Communication links

Scientists on Earth stay in touch with spacecraft through networks of antennae positioned around the planet. The USA's Deep Space Network has satellite dishes at three sites in Australia, Spain, and the USA – these work together to provide constant contact with spacecraft as Earth rotates.

In-depth orbits

Spacecraft have orbited six of the Solar System planets, from Mercury out to Saturn. By circling these worlds, they can make systematic studies of them. Whole planets can be mapped – and changes recorded – on a daily, monthly, or yearly basis. Juno (right) arrived at Jupiter in 2016 and moved into a polar orbit to start its year-long study.

Three solar panels around Juno's hexagonal body provide electrical power.

The Sun

Our local star generates vast amounts of energy, which is released into space and fuels life on Earth. Spacecraft monitor activity – from sunspots to huge jets of gas – and help forecast its effect on Earth. In the future, the Sun will dramatically expand in size, destroying Earth.

Our local star

The largest and most massive body in the Solar System, the Sun accounts for more than 99 per cent of the system's mass. It is a ball of hot gas – mostly hydrogen and helium – with no solid surface. In its core, nuclear fusion converts hydrogen to helium, producing energy that is most familiar to us as heat and light.

The Sun's influence on life

The Sun warms Earth and powers the water cycle (see p.24). Through photosynthesis, plants convert sunlight into chemical energy, as carbohydrates. These nourish animals that eat the plants. Photosynthesis also releases oxygen, which is vital to life.

Huge solar eruptions

Studying the Sun

The Sun is best observed from space. The Solar and Heliospheric Observatory (SOHO), positioned 1.5 million km (0.9 million miles) from Earth, has been watching the Sun continuously since 1996. It has observed two sunspot cycles and studied more than 20,000 coronal mass ejections.

The four solar panels that power SOHO span about 7.6 m (25 ft) in total.

Solar wind

The Sun constantly emits a stream of tiny particles called the solar wind, which shapes the magnetic field around planets and causes auroras. These spectacular light displays (above) occur when solar wind particles are drawn into the atmosphere above Earth's poles, collide with atoms, and cause them to give off light.

Sunspots

Dark, temporary patches on the Sun's surface called sunspots are cooler regions where the Sun's magnetic field interrupts rising heat. The number of spots varies in an 11-year cycle. Around 2020 will be a time of least spots, called solar minimum. The time of maximum spots is associated with coronal mass ejections.

A large sunspot seen in October 2014 could fit 10 Earths across it.

The Little Ice Age

The prolonged period of few sunspots called the Maunder minimum in 1645–1715 was a time of particularly cold weather known as the Little Ice Age. If and how the two are related is uncertain. Rivers that normally flow freely all year froze over (right).

Coronal mass ejections

Made of billions of tonnes of gas, a coronal mass ejection speeds away from the Sun (above). If it travels in Earth's direction, it can cause a solar storm. Solar storms can damage satellites, disrupt communications, cause surges in power lines, and trigger an aurora.

Solar eclipse

The Sun's atmosphere extends far beyond its visible surface. Normally invisible, the outermost corona can be seen when the Sun is eclipsed (left). A total solar eclipse occurs when the Moon is directly between the Sun and Earth, covering the Sun's face.

In the future

In about 5 billion years, the Sun will run low on hydrogen and expand, engulfing Mercury, Venus, and Earth. Before then, its intense heat will kill all life, evaporate oceans (above), and turn Earth molten.

Mercury

Sun-baked Mercury is the closest planet to the Sun. Here, the Sun shines seven times hotter than on Earth. But Mercury's thin atmosphere can't retain the heat, so it is freezing cold at night. Mercury has a huge iron core and a rocky, barren surface covered in craters.

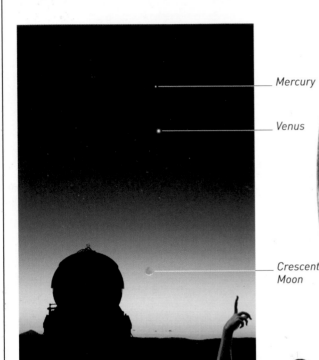

Mercury

Venus

Crescent Moon

View from Earth

Mercury is never far from the Sun in our sky and is usually hidden by the Sun's light. But six times a year – during spring and autumn – we see Mercury well when its orbit brings it into view first on one side of the Sun and then the other for about two weeks at a time.

Fast mover

Named after the swift-footed messenger of the Roman gods, Mercury speeds around the Sun faster than any other planet. It takes just 88 days to complete one orbit. Yet Mercury spins on its axis very slowly, rotating once in just under 59 days, and completes just three spins in two orbits. The combination of slow spin and fast orbit means there are 176 days between one sunrise and the next.

The Amaral Crater was named after Brazilian artist Tarsila do Amaral.

Rocky surface

Mercury's rocky surface has hardly changed over the past 3 billion years. The impact craters found all over the planet formed when asteroids smashed into young Mercury. Its smooth plains formed when volcanic lava flowed over parts of its surface. Long ridges and cliffs are the result of Mercury's surface crust shrinking unevenly as the young planet cooled.

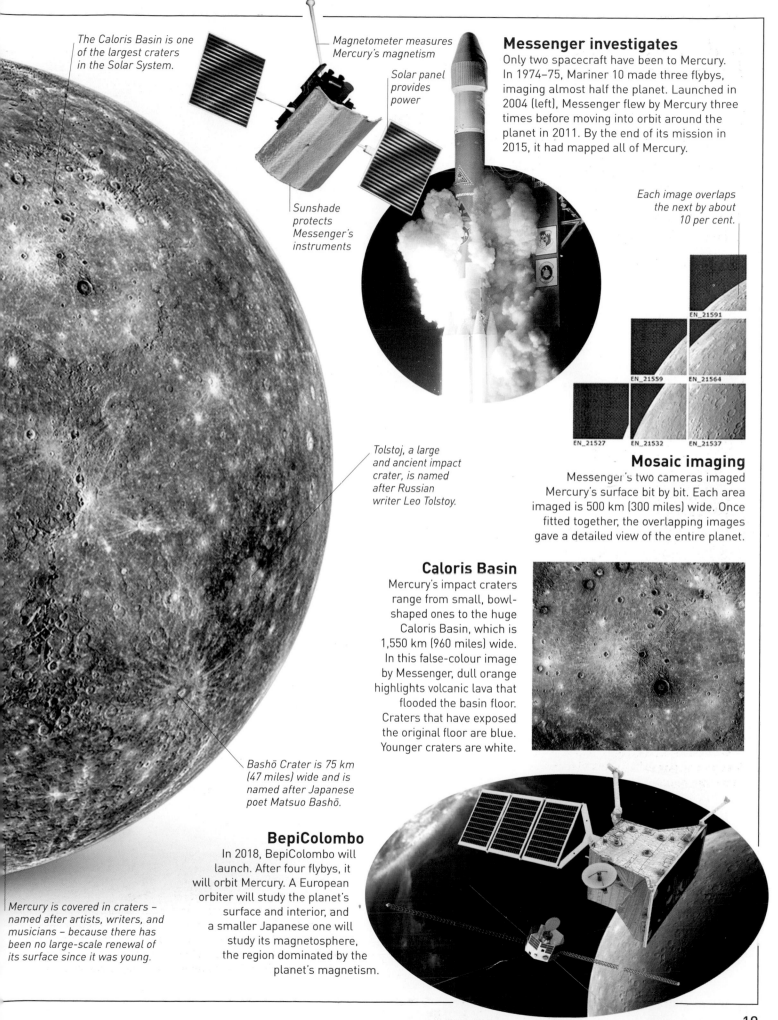

The Caloris Basin is one of the largest craters in the Solar System.

Magnetometer measures Mercury's magnetism

Solar panel provides power

Sunshade protects Messenger's instruments

Messenger investigates

Only two spacecraft have been to Mercury. In 1974–75, Mariner 10 made three flybys, imaging almost half the planet. Launched in 2004 (left), Messenger flew by Mercury three times before moving into orbit around the planet in 2011. By the end of its mission in 2015, it had mapped all of Mercury.

Each image overlaps the next by about 10 per cent.

EN_21591

EN_21559   EN_21564

EN_21527   EN_21532   EN_21537

Mosaic imaging

Messenger's two cameras imaged Mercury's surface bit by bit. Each area imaged is 500 km (300 miles) wide. Once fitted together, the overlapping images gave a detailed view of the entire planet.

Tolstoj, a large and ancient impact crater, is named after Russian writer Leo Tolstoy.

Caloris Basin

Mercury's impact craters range from small, bowl-shaped ones to the huge Caloris Basin, which is 1,550 km (960 miles) wide. In this false-colour image by Messenger, dull orange highlights volcanic lava that flooded the basin floor. Craters that have exposed the original floor are blue. Younger craters are white.

Bashō Crater is 75 km (47 miles) wide and is named after Japanese poet Matsuo Bashō.

BepiColombo

In 2018, BepiColombo will launch. After four flybys, it will orbit Mercury. A European orbiter will study the planet's surface and interior, and a smaller Japanese one will study its magnetosphere, the region dominated by the planet's magnetism.

Mercury is covered in craters – named after artists, writers, and musicians – because there has been no large-scale renewal of its surface since it was young.

Venus

Almost as big as Earth, Venus is twice as far from the Sun as Mercury, but its surface temperature of 464°C (867°F) makes it the hottest planet of all. Its suffocating atmosphere hides a landscape of lava plains and volcanoes.

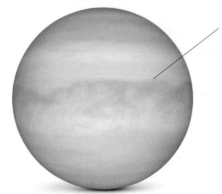

Venus has the thickest and densest atmosphere of all the rocky planets.

Atmosphere

Venus's carbon-dioxide atmosphere extends about 80 km (50 miles) above the surface. Its layers of cloud contain sulphuric acid droplets. Venus takes eight months to rotate, but the high clouds speed round Venus in just four days.

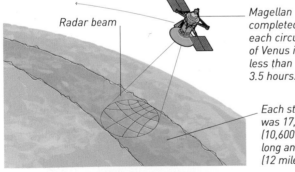

Radar beam

Magellan completed each circuit of Venus in less than 3.5 hours.

Each strip imaged was 17,000 km (10,600 miles) long and 20 km (12 miles) wide.

Looking through the clouds

Orbiting Venus in 1990–94, Magellan used radar pulses to peer through the clouds. The signals that bounced back from the surface built up images in strips, which once combined, gave us our current global view.

Dali Chasma is a system of canyons and troughs more than 2,000 km (1,240 miles) long.

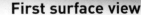

First surface view

The Venera craft were the first to land on Venus's surface. Venera 9 sent back a black-and-white view in 1975. In 1982, Venera 13 took the first colour images (left).

Woman's world

This radar view by Magellan shows Venus's rocky surface, with huge volcanoes, lava plains, canyons, and impact craters formed in the last 500 million years, after the main era of volcanism. Venus is named after the Roman goddess of love. All but one of its surface features have women's names.

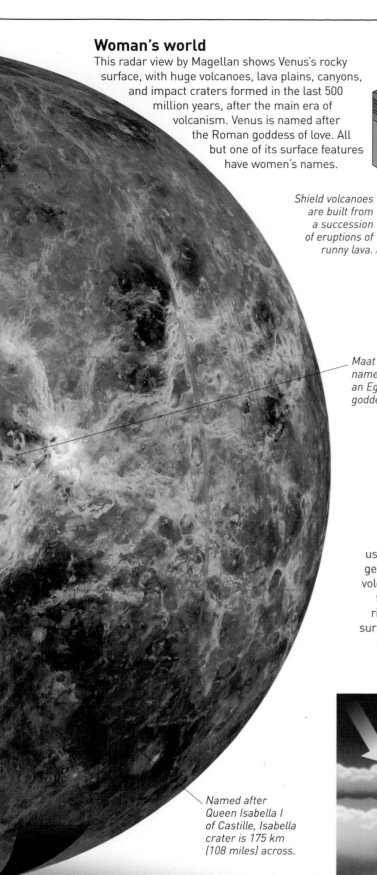

Shield volcanoes are built from a succession of eruptions of runny lava.

A pancake dome forms when thick lava erupts slowly.

An arachnoid volcano has fracture lines radiating out from its centre.

Volcanic surface

More than 85 per cent of Venus is covered in volcanic lava. Hundreds of the volcanoes that released the lava are shield volcanoes, with gentle slopes built up by successive lava flows. The smaller arachnoid volcanoes, named for their spider-web-like appearance, and the flat-topped pancake domes are unique to Venus. Most of the volcanoes are thought to be extinct.

Maat Mons is named after an Egyptian goddess.

The caldera (crater) is about 30 km (20 miles) wide at the summit.

Maat Mons

Magellan data has been used to create this computer-generated image of the shield volcano Maat Mons (right). It is the tallest volcano on Venus, rising 8 km (5 miles) above the surrounding landscape. Lava flows extend for hundreds of kilometres across the plains in the foreground.

Named after Queen Isabella I of Castille, Isabella crater is 175 km (108 miles) across.

Greenhouse effect

Most of the sunlight reaching Venus is reflected back into space by the top layer of its clouds, making the planet bright and easy to see from Earth. A small amount of sunlight gets through the clouds, warming the rocky surface. Heat released from the rock adds to the warming process. The clouds work like the glass in a greenhouse, keeping heat in.

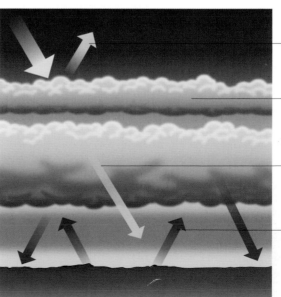

About 80 per cent of sunlight is reflected off Venus's clouds.

The thick layers of cloud stop heat from Venus escaping into space.

About 20 per cent of the sunlight reaches Venus's surface.

Radiation from the Sun-warmed ground is absorbed by carbon dioxide in Venus's atmosphere and cannot escape into space.

Earth

Our home planet is unique for its liquid water oceans and the presence of life. The largest of the four inner planets, Earth is made of rock and metal, and gets denser and hotter towards its core. Surrounding it is a nitrogen-rich atmosphere protecting life below.

Earth in space
The crew of Apollo 17 were heading for the Moon on 7 December 1972 when they captured this view of Earth as a blue-and-white marble. Wispy white clouds overlay deep blue oceans and the pale brown continents of land that cover more than a quarter of Earth's surface.

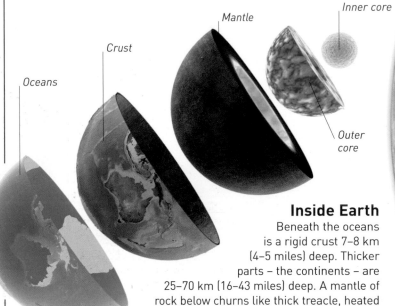

Oceans

Crust

Mantle

Inner core

Outer core

Inside Earth
Beneath the oceans is a rigid crust 7–8 km (4–5 miles) deep. Thicker parts – the continents – are 25–70 km (16–43 miles) deep. A mantle of rock below churns like thick treacle, heated by the iron and nickel core. The outer core is molten, but the inner core, though hotter, is solid.

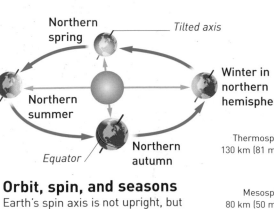

Northern spring

Tilted axis

Winter in northern hemisphere

Northern summer

Equator

Northern autumn

Orbit, spin, and seasons
Earth's spin axis is not upright, but tilted at an angle of 23.4 degrees. This causes the seasons. When one hemisphere points to the Sun, it has the warm, long days of summer, while the other hemisphere points away and experiences winter. Six months later, the seasons reverse.

Thermosphere
130 km (81 miles)

Mesosphere
80 km (50 miles)

Ozone layer

Stratosphere
50 km (31 miles)

Troposphere
10 km (6 miles)

Aurora

Meteor streaks through the atmosphere

Atmosphere
Earth's atmosphere is mainly nitrogen and oxygen. Its layers are shown here with their upper heights. Clouds and weather occur in the troposphere, which contains 90 per cent of the atmosphere's gas and is the only part with breathable air. In the stratosphere, the ozone layer absorbs harmful radiation from the Sun. Beyond this, the atmosphere thins until it merges with space.

Tectonic plates

Earth's crust is joined to the top of the mantle, in the lithosphere. This is split into tectonic plates – seven large ones (left) and many smaller ones. Lying on top of the semi-molten mantle, the plates move at about the same speed that fingernails grow.

Eurasian plate
North American plate
Pacific plate
South American plate

Antarctic plate
African plate
Australian plate

Volcanic eruptions create new land through ash and lava deposits.

Snow and rain feed glaciers and streams, which erode rocks.

The remains of marine life and rock pieces compact to form sedimentary rock.

One plate under another causes a volcanic mountain range to form.

Metamorphic rock forms deep inside Earth as a result of heat and pressure.

The rock cycle

On Earth's ever-changing surface, rocks are broken down by water and weather, and pieces are moved by glaciers, wind, and water. Those carried to the sea compact into layers of rock. Tectonic activity brings rock to the surface as mid-ocean island chains, mountains, or lava.

The continent of Antarctica is almost entirely covered in ice.

Roof of the world

Where tectonic plates push together and buckle, mountains grow as rock layers stack on top of one another. The Himalayas (right) – Earth's highest mountains – took shape in the last 50 million years. The range is rising by up to 4 mm (0.2 in) a year, offset by weathering and erosion.

Ring of fire

More than three-quarters of land volcanoes occur where two plates converge – many in the Ring of Fire, an arc of volcanic and earthquake activity around the Pacific Ocean, where smaller plates meet the Pacific plate. Its volcanoes include the Philippines' Mount Pinatubo (above).

Water world

Earth's water makes our planet unique. Oceans and seas of liquid salty water cover about 70 per cent of Earth's surface. Fresh water frozen in glaciers, ice sheets, and icebergs brings the total to more than 80 per cent. Never still, they and Earth's rivers display water's power.

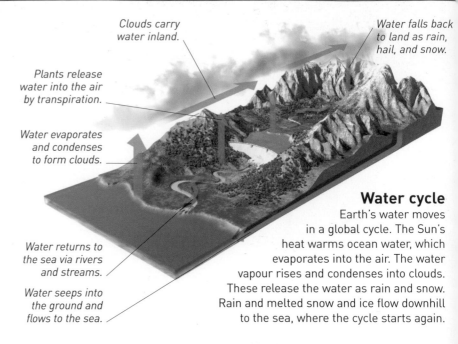

Clouds carry water inland.

Water falls back to land as rain, hail, and snow.

Plants release water into the air by transpiration.

Water evaporates and condenses to form clouds.

Water returns to the sea via rivers and streams.

Water seeps into the ground and flows to the sea.

Water cycle
Earth's water moves in a global cycle. The Sun's heat warms ocean water, which evaporates into the air. The water vapour rises and condenses into clouds. These release the water as rain and snow. Rain and melted snow and ice flow downhill to the sea, where the cycle starts again.

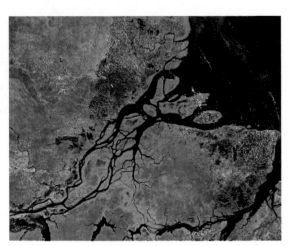

Amazon River
Rivers hold less than one per cent of Earth's water, but have a big effect on its landscape, carrying about 20 billion tonnes of sediment to the oceans every year. The Amazon – whose mouths are seen here from space – is 6,430 km (3,995 miles) long and delivers a fifth of all river water reaching the sea.

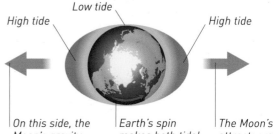

Low tide

High tide

High tide

On this side, the Moon's gravity attracts Earth more than water.

Earth's spin makes both tidal bulges sweep over the surface.

The Moon's gravity attracts water more than Earth.

Daily tides
The Moon's gravity pulls on the oceans. The pull is stronger nearer to the Moon, so a bulge of water forms on the side nearest to the Moon, and on the opposite side. As Earth turns, the bulges create daily changes in the sea level – our high and low tides.

Blue planet
The largest of the five oceans – the Pacific – covers more than a third of Earth's surface and holds more than half of its liquid water. Next is the Atlantic, and then the Indian Ocean. The two smallest – the Arctic and the Southern Ocean – are in the polar regions and have huge amounts of ice floating in them.

Frozen water

More than three-quarters of Earth's fresh water is ice – in glaciers, ice sheets and shelves (above), icebergs, mountain-top coverings, and soil. Most of it is in the ice sheet covering Antarctica – if it melted, sea levels would rise by about 60 m (197 ft).

Jason-3 orbits 1,336 m (830 miles) above Earth, passing over the same point every 10 days.

A radar altimeter measures wave height and wind speed.

The Hawaiian Islands rise 9.5 km (5.9 miles) from the ocean floor.

Hawaii, the largest island in the chain, is five merged volcanoes.

Under the oceans

The ocean floor is mostly flat plains, but it also has mountains and trenches. The Mariana Trench plunges 11 km (6.8 miles) below the Pacific Ocean's surface. The Mid-Atlantic Ridge is Earth's longest mountain range. Deep-sea volcanoes that break through the water's surface make islands such as Hawaii.

Water watch

Satellites orbiting Earth monitor its land, oceans, and ice. Jason-3 (above) measures the height of the ocean surface as part of a wider study of changes in sea levels and the effects of climate change. The Aqua satellite studies the water cycle, and CryoSat measures changes in the thickness of the ice sheets.

The Pacific Ocean covers 156 million km² (60 million sq miles).

Living planet

Life began at least 3.7 billion years ago, and evolved from small bacteria-like cells to the huge variety we see today. Human expansion across the planet has contributed to the extinction of other life forms, and to climate change. Yet we are also fascinated by the potential for life on other planets.

Myriad life forms

Earth teems with about 8.7 million species. Each species has its own set of characteristics that suit its own particular environment. There are almost 1 million species of insects, including around 40 species of leafcutter ants (above).

Beginnings

Life developed from very basic, self-replicating molecules into primitive cells. Multi-cellular organisms gave rise to more complex life such as plants and animals. Stromatolites (left) are layered-rock structures formed by single-cell microbes – similar ones existed on Earth 3.5 billion years ago.

Kingdoms of life

From tiny micro-organisms visible only through a microscope to Earth's largest animals – such as the African elephant (below) – scientists group the huge variety of life forms into five kingdoms: animals, plants, fungi, protists, and monerans. Monerans are single cells with no internal structure. The protists are single cells but with a nucleus inside. Fungi include yeasts and mushrooms. The most complex life forms are plants and animals.

Elephants have existed for around 55 million years, but are now vulnerable to habitat loss and the illegal trade in their ivory tusks.

Human impact

Global population growth has led to cutting down forests for farmland, destroying the habitat of thousands of species and reducing the conversion of carbon dioxide to oxygen by trees. As a result, the temperature on Earth increases, polar ice caps melt, and sea levels rise.

Lonesome George, the last Pinta Island tortoise, died on 24 June 2012.

Tokyo, Japan, is one of Earth's largest and most populated cities

Ozone layer

The ozone layer in Earth's atmosphere blocks the Sun's ultraviolet light, which can damage living cells. Scientists in the 1970s found holes in the ozone layer – this 2014 image shows the hole over Antarctica in blue. The chemicals causing the damage have now been banned in almost 200 countries, and the hole is shrinking.

Extinction

Extinction is a natural part of the evolution of life. Some species adapt and survive events such as climate change or competition from other species, but many die out and become extinct. Many more species are extinct than are alive today.

Part of the USA's Allen Telescope Array of more than 40 radio telescopes listening for signals from extra-terrestrial life

Extremophiles

Micro-organisms called extremophiles thrive in extreme environments such as scalding and acidic water. Fringing the hot waters of Grand Prismatic Spring (right) in Yellowstone, USA, are orange mats of microbes called archaea, which survive temperatures of up to 74°C (165°F).

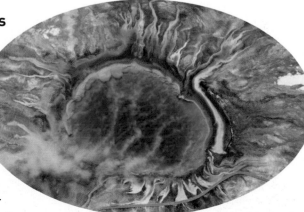

Search for life elsewhere

No evidence of life beyond Earth has yet been found, but spacecraft are searching for life on Mars, and Jupiter's moon Europa is the next target. Beyond the Solar System, exoplanets in the habitable zone of their star might harbour life.

The Moon

The Moon is Earth's nearest neighbour, a quarter of Earth's size, and the largest and brightest object in our night sky. Yet this ball of rock and metal has a dry, dead surface, with huge volcanic plains and highlands covered by impact craters. As it orbits Earth, it appears to change shape, from a slim crescent to a full Moon.

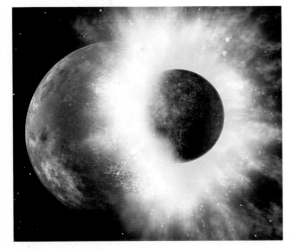

Violent origin

About 4.5 billion years ago, a Mars-sized asteroid gave young Earth a glancing blow. Material from both splashed into space, forming a ring around Earth, and then slowly clumped together to form the young Moon.

Lower mantle of partially molten rock

Mantle of solid rock

Crust of solid rock

Inner core of solid iron

Outer core of molten iron

Inside the Moon

Once the Moon had formed a ball, it began to cool. Its heavy metals sank as rocks formed its outer layers. Today, the interior has largely solidified around a core of hot iron. The inner core is about 1,400°C (2,600°F), but it is squeezed solid by the rocks around it.

Familiar face

The Moon shines by reflected sunlight, and its surface features are easily seen with the naked eye. The near-side face (above) is always turned towards us. Its dark patches are areas of volcanic lava called maria. The lighter areas are older mountainous regions shaped by asteroid impact.

Tycho Crater is about 110 million years old and 85 km (53 miles) across.

Waxing crescent

First quarter

Waxing gibbous

Full Moon

Waning gibbous

Last quarter

Waning crescent

New Moon

Same face always points to Earth

Orbit and rotation

The Moon takes 27.3 days to orbit Earth. As it orbits, the Moon also spins, taking 27.3 days for one rotation – the same time as one orbit. As a result, the same side of the Moon, the near side, shown here in green, always faces Earth.

Day 21

Day 1

Earth

Day 14

Day 7

Direction of Moon's orbit

Ancient landscape

The young Moon was bombarded by asteroids, forming craters and pushing up mountains. Volcanism then formed the maria as lava flooded many craters. Little has changed for two billion years except the formation of craters such as Tycho (above).

Lunar eclipse

When Earth is directly between the Sun and the Moon, it blocks the sunlight. The Moon is in Earth's shadow, and is eclipsed. A small amount of sunlight passing around Earth reaches the Moon and makes it look reddish.

Moon myths

Across different cultures, there are centuries-old stories of the full Moon causing fevers and diseases, turning humans mad, or changing them into werewolves. The word lunatic comes directly from *Luna*, the Latin word for Moon.

Lunar phases

The Moon's shape (or phase) appears to change day by day, according to how much of the near side is lit by sunlight. When the Moon is between Earth and the Sun, its near side is dark (new Moon). As it orbits, the Moon appears to grow (waxes) until fully lit (full Moon) and then shrinks (wanes).

Exploring the Moon

People have studied the Moon for thousands of years. Since 1959, more than 60 spacecraft have successfully travelled there. Six Apollo missions took 12 men to its surface between 1969 and 1972. It remains the only place we have visited beyond Earth.

Mare Moscoviense (Sea of Muscovy) is 276 km (17 miles) wide and the second largest mare on the far side.

Far side of the Moon

In 1959, the Luna 3 spacecraft took the first images of the far side of the Moon. Only the Apollo astronauts who have orbited the Moon have seen that side in person. It is more heavily cratered, and because it had less volcanic activity, it has no large maria.

Galileo and the Moon

Italy's Galileo Galilei was the first to study the Moon through a telescope. In 1609, he saw that it is not flat, as previously thought, but has mountains, craters, and smoother dark areas. His observations and sketches (such as the lunar phases above) became widely known.

Men on the Moon

In the 1960s, American spacecraft were sent to the Moon to photograph the surface and soft-land on it. Once a site was chosen for a manned landing, Apollo 11 and its crew blasted off for the Moon. Neil Armstrong (far left) and Buzz Aldrin (right) took the first ever footsteps on the Moon, on 21 July 1969, as Michael Collins (centre) orbited overhead.

Rock and soil

The Apollo astronauts collected 2,200 rock and soil samples weighing a total of 382 kg (842 lb). The breccia rocks formed when asteroid impacts melted and compacted the surface rock and soil. The basalts are volcanic rock formed from lava that seeped through the Moon's crust, creating the maria.

Breccia rock from the Apollo 17 mission

The rover travelled at up to 18.5 kph (11.5 mph).

Driving around

The last three Apollo missions – Apollo 15, 16, and 17 – used a Lunar Roving Vehicle to explore a wider area. It could carry two astronauts, cameras, tools, and rock and soil samples.

Apollo 17's Eugene Cernan

Lunar Reconnaissance Orbiter

Return to the Moon

Many spacecraft were sent to the Moon in the 1960s and 70s, but none in the 1980s. Recent craft such as the Lunar Reconnaissance Orbiter – orbiting the Moon since 1990 – carry out scientific research. Japan, Europe, China, and India have sent craft, and some of them aim to be the first to return humans to the Moon.

Instrument looks for evidence of water as ice

Luna landings

The Soviet Union did not land a man on the Moon, but it did launch key missions. Luna 9 made the first soft landing there, in 1966, proving that the lunar soil could support landing craft. Luna 16 (featured on stamps, above) was the first to return a soil sample to Earth, in 1970.

Apollo 17 geologist Harrison Schmitt returning to the rover.

Walking on the Moon

The 12 men who walked on the Moon explored six sites. As the Moon has one sixth of Earth's gravity, they had to push off with one foot and float forward before planting the next. The final mission, Apollo 17, explored the Moon's Taurus-Littrow Valley.

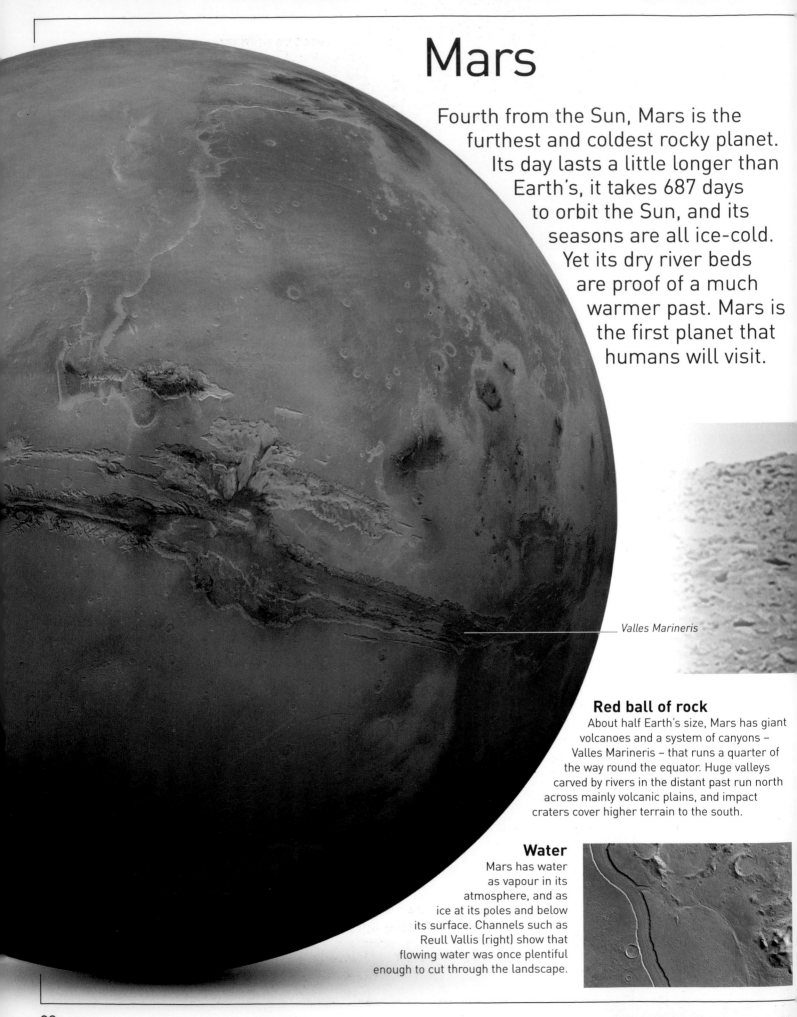

Mars

Fourth from the Sun, Mars is the furthest and coldest rocky planet. Its day lasts a little longer than Earth's, it takes 687 days to orbit the Sun, and its seasons are all ice-cold. Yet its dry river beds are proof of a much warmer past. Mars is the first planet that humans will visit.

Valles Marineris

Red ball of rock

About half Earth's size, Mars has giant volcanoes and a system of canyons – Valles Marineris – that runs a quarter of the way round the equator. Huge valleys carved by rivers in the distant past run north across mainly volcanic plains, and impact craters cover higher terrain to the south.

Water

Mars has water as vapour in its atmosphere, and as ice at its poles and below its surface. Channels such as Reull Vallis (right) show that flowing water was once plentiful enough to cut through the landscape.

Phobos is 26.8 km (16.7 miles) long, cratered, and covered in loose rock and dust.

Deimos

Phobos

Moons of Mars

Mars has two small, irregular-shaped, rocky moons, named after two sons of Ares, the Greek god of war. At a distance of 9,378 km (5,827 miles) from Mars, Phobos is closer to its planet than any other moon, orbiting Mars in just over 7.5 hours. Smaller Deimos is twice as far away and its orbit takes four times as long. Their origin is uncertain. One idea is that they were asteroids captured by Mars's gravity. Another is that the two formed from debris after a huge asteroid collided with young Mars.

A model of the Viking 1 lander, which tested its surroundings for signs of life on Mars, with inconclusive results

First close-up

Two identical Viking missions arrived at Mars in 1976. Viking 1 transmitted the first black-and-white image from the Martian surface, followed by the first colour image (below), revealing a red-coloured rocky terrain with fine-grained sand under a red sky.

Polar caps

Like Earth, Mars has a white ice cap at each pole. These huge mounds tower over the surrounding landscape. The northern cap is frozen water; the southern is water ice topped by a permanent layer of carbon-dioxide ice. Both caps extend in the winter and shrink in the warmer summer.

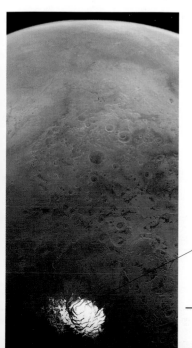

Mars's southern cap, named Planum Australe, is about 420 km (260 miles) across in summer.

Putting people on Mars

Space agencies and private companies aim to take people to Mars by about 2050. Simulated missions on Hawaii (above) keep crews in isolation to see how they might cope on a round trip up to three years long.

The Red Planet

Mars is a cold, desert-like world with rock-strewn plains, hills, huge shield volcanoes, and vast canyons. Its rocky crust is in one solid piece over a mantle of rock. When Mars was young, the mantle was warmer and fluid. Its movement tore the crust apart, forming the largest canyons and volcanoes in the Solar System.

Blood-red world

Mars's blood-red appearance in our night sky led to it being named after the Roman god of war. The colour comes from iron oxide (rust) in the soil that covers almost the entire globe – winds sweep rusty particles up into the thin carbon-dioxide atmosphere.

Olympus Mons

Mars's biggest volcanoes are in the Tharsis Bulge, west of the Valles Marineris. The largest of all, Olympus Mons is 22 km (14 miles) high – almost three times the height of Earth's Mount Everest – and almost as wide as Germany. This huge shield volcano grew gradually as lava flows built up over millions of years.

Land has slid down the slopes and collected in the canyon floor.

Melas Chasma is the central and widest part of Valles Marineris.

Red sky, blue sunset

Mars's sky is red because of dust particles in the atmosphere. At sunset, the sky around the setting Sun turns blue as dust high in the atmosphere scatters sunlight. The rover Spirit captured this sunset in 2005. The Sun is white and smaller than in our sky because Mars is one and a half times further from the Sun than Earth.

The blue glow lasts for more than two hours after the Sun sets below the horizon.

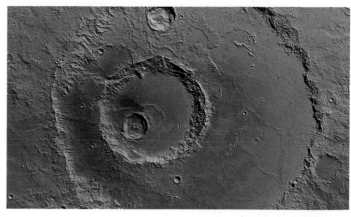

Craters within craters

Impact craters are found all over Mars's surface, but especially in its southern regions. Many to the north were covered by lava flows. The largest in this image is Hadley Crater, which is 120 km (75 miles) wide. Lava has flooded its floor, but later impacts each blasted deeper into the surface.

Valles Marineris

This gigantic system of canyons is 4,000 km (2,500 miles) long and, on average, 8 km (5 miles) deep. Five times deeper and nearly ten times longer than Earth's Grand Canyon, the Valles Marineris would stretch across North America. It formed when the Tharsis Bulge swelled with magma during the planet's first billion years, the nearby crust stretched and split, and land between the cracks dropped.

Fields of dunes

Sand dunes are common on Mars. They formed as wind piled up the sand and then sculpted it into different shapes. Some look like Earth's dunes, but others have less familiar shapes, such as long straight lines formed by wind consistently blowing in one direction. The crests of these dunes in Endurance Crater are less than 1 m (3 ft) high.

Roving on Mars

Of more than 20 successful missions to Mars, seven landed on its surface. Three stayed put, and four roved over the planet's rust-red landscape. Operated by on-board computers directed from Earth, and fitted with cameras and rock analysis tools, the rovers record their findings and send them back to Earth.

Landing on Mars

Curiosity took just over eight months to travel to Mars, packed inside a protective casing. Lowered to the surface by a descent stage (above), it then cut itself free and the stage flew clear. Previous rovers bounced on to the surface inside a ball of airbags, and then drove out.

The rovers

Sojourner was the first rover on Mars. The size of a microwave oven, it worked near its landing site for three months in 1997. The more advanced twin rovers, Spirit and Opportunity, arrived on opposite sides of Mars in 2004. Opportunity continues to explore, but Spirit no longer works. Curiosity has been roving since it arrived in 2012.

SOJOURNER
July–September 1997
Distance travelled: 100 m (330 ft)

SPIRIT
January 2004–March 2010
Distance travelled: 7.7 km (4.8 miles)

OPPORTUNITY
January 2004–present
Distance travelled: 43 km (26.7 miles)

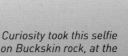

CURIOSITY
August 2012–present
Distance travelled: 13.7 km (8.5 miles)

Curiosity

The latest, largest, and most sophisticated rover is Curiosity. This car-sized robot is studying Mars's rocks and soil to learn about its climate and geology, and to find out if it could have supported life in the past. Curiosity's self-portrait on 5 August 2015 combines several images taken by one of its 17 cameras.

Curiosity took this selfie on Buckskin rock, at the base of Mount Sharp in the centre of Gale Crater.

Curiosity's ChemCam

The ChemCam instrument on Curiosity looks at rocks and soil from up to 7 m (23 ft) away. Mounted on a mast, it uses a laser to vaporize the target rock surface, and a camera takes images. Another instrument analyses light energy emitted by the rock and determines the rock's composition.

Drilling into Mars

Spirit, Opportunity, and Curiosity have all drilled holes into Mars to analyse samples from inside its rock. Curiosity first used the drill on the end of its robotic arm in 2013. This produced a powdered sample, which the rover's analytical instruments found showed evidence of an ancient wet environment.

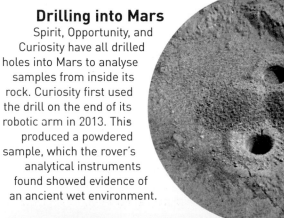

Staying in touch

Curiosity sends data directly to Earth or via a spacecraft orbiting Mars – either the Mars Reconnaissance Orbiter (above) or Mars Odyssey. The messages are received by the Deep Space Network's dish antennas in the USA, Spain, and Australia.

The ExoMars Trace Gas Orbiter will analyse Mars's atmosphere and relay data from future rovers.

Schiaparelli

ChemCam's laser emits pulses from here.

The robotic arm, bearing the camera that captured this selfie, is not shown in full in this composite image.

Future rovers

New Martian rovers are already in preparation. A small lander called Schiaparelli was released in 2016 by the ExoMars Trace Gas Orbiter to test the descent and landing procedure for a European rover scheduled for touchdown in 2020. The USA's Mars 2020 will arrive at about the same time. It will collect rock samples for a future mission to pick up and return to Earth.

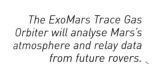

Investigating Vesta

Vesta is the brightest asteroid in our sky and the only one visible to the naked eye from Earth. Images taken by the spacecraft Dawn in 2011–12 revealed the scars of collisions with other asteroids. One gouged out a large impact crater near the south pole and sent pieces of Vesta flying. Some landed on Earth, where they were collected as meteorites.

Craters cover Vesta in this view of its northern hemisphere, taken by Dawn in August 2012.

Meteorite evidence

This false-colour microscopic view shows various minerals in a 6-mm- (0.2-in-) wide slice of a meteorite thought to have come from Vesta.

Asteroids

There are millions of asteroids in the Solar System, with most in the Asteroid Belt between Mars and Jupiter. Left over from when the planets were forming, these small, rocky bodies follow their own orbits around the Sun. The first asteroid to be discovered was Ceres, in 1801. We have identified almost 600,000 asteroids and the number is rising.

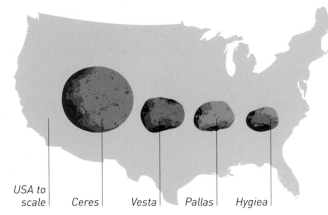

USA to scale   *Ceres*   *Vesta*   *Pallas*   *Hygiea*

Size and shape

Ceres is 952 km (592 miles) across. As the largest, roundest asteroid, it is also classed as a dwarf planet. Next in size are Vesta, Pallas, and Hygiea. Only 26 asteroids are bigger than 200 km (124 miles) across. Others are irregular in shape and just a few kilometres wide.

The Asteroid Belt

The asteroids in this doughnut-shaped ring typically take four to five years to orbit the Sun. Once the orbit of a newly found asteroid is known, the body may be assigned a name chosen by the discoverer. There are also two swarms of asteroids that have similar orbits to Jupiter, and are named the Trojans.

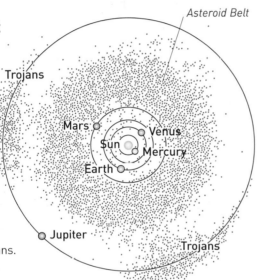

Asteroid Belt

Trojans

Mars
Venus
Sun
Mercury
Earth

Jupiter

Trojans

Close encounters

Asteroids that come within about 7.5 million km (4.7 million miles) of our planet are potential threats. Hundreds of the larger ones are closely monitored, but small ones arrive unexpectedly. In 2013, an asteroid 18 m (59 ft) wide broke up above Chelyabinsk, Russia, creating a trail of gas and dust (above). Larger chunks fell to the ground as meteorites.

Impact with Earth

Large asteroid impacts on Earth are very rare. An asteroid at least 150 m (492 ft) wide strikes Earth roughly every 10,000 years and would produce a crater about 2.5 km (1.5 miles) wide. One of Earth's 190 impact craters, Pingualuit Crater in Canada (above), is the result of an impact around 1.4 million years ago. It is 3.4 km (2.1 miles) across.

Down to Earth

Tonnes of asteroid material reach Earth every day. Much of it is in the form of small pieces that burn up in the atmosphere. Pieces too large to burn up reach the surface and are called meteorites. Each year, around 3,000 meteorites weighing more than 1 kg (2.2 lbs) land on Earth.

A meteorite weighing 2,045 kg (2.25 tons) was found in Saudi Arabia's Empty Quarter in 1966.

Studied by spacecraft

The first dedicated mission to an asteroid was NEAR Shoemaker, which landed on Eros in 2001. Hayabusa (left) collected dust from Itokawa in 2005 and sent it back to Earth in a capsule that landed in Australia in 2010.

Jupiter

The Solar System's largest and most massive planet is named after the king of the Roman gods and ruler of the sky. Five times the distance of Earth from the Sun, and the closest giant planet, Jupiter is made of mainly hydrogen and helium, with no solid surface. When we look at it, we see only the top of its deep atmosphere.

Giant world

Jupiter is made of almost two and a half times the material of the other seven planets combined, and 11 Earths would fit across its face. Yet it is the fastest spinner of all the planets, turning once in less than 10 hours. Its rapid rotation, fierce winds, and the large-scale movement of gases create the stripes that run parallel to its equator.

The Great Red Spot is the largest storm in the Solar System.

Comet crash

Jupiter's huge gravitational pull can change the orbit of a passing asteroid, distort a moon's shape, or break up a comet. When comet Shoemaker-Levy was pulled into orbit around Jupiter and broke up, 21 pieces stretching across 1.1 million km (710,000 miles) were seen on a collision course with Jupiter, crashing in July 1994.

Coloured clouds

Hydrogen gas dominates Jupiter's atmosphere. The rest is mostly helium, with small amounts of hydrogen compounds that form the colourful bands of clouds at different heights. Warm gases rising from the lower, darker bands (called belts) form high, icy clouds in the paler bands (zones), then cool and sink in a constant cycle.

Winds sweep high clouds into bands.

High clouds are colder and paler.

Low clouds are warmer and darker.

Warm gases rise, cool, and then sink.

Water clouds form deeper in the atmosphere.

Jupiter's auroras are hundreds of times brighter than Earth's.

Magnetic field

Jupiter has a magnetic field generated by electric currents within its body. This field influences a huge region of space around the planet. Solar wind particles and others from Jupiter's moon Io are drawn into the upper atmosphere above the poles, creating light displays called auroras.

Atmosphere

Winds in this dark belt stream eastwards at 480 km (298 miles) per hour.

Liquid-like layer

Liquid metallic hydrogen layer

Solid core

Inside Jupiter

Jupiter's hydrogen and helium are in a gaseous form in the outer atmosphere. Inside, as pressure, density, and temperature increase, they become more liquid-like. Deeper still, the hydrogen swirls like a liquid metal. At the centre is a rocky core.

Ring discovery

Jupiter's rings – discovered in 1979 by Voyager 1 – are made of fine, dark dust grains from the surface of nearby moons. The grains are so thinly spread that the rings are almost transparent.

The Great Red Spot rotates anticlockwise once in six days.

This false-colour image shows high clouds in pink, low clouds in blue, and the deepest ones in black.

Stormy surface

In the upper atmosphere, Jupiter's non-stop winds, fast spin, and internal heat stir up giant storms that we see as ovals. Some are short-lived; others last for decades. The Great Red Spot was once three times the size of Earth, but it would fit just one Earth inside it now.

Jupiter's moons

Jupiter is orbited by at least 67 moons – the largest family of moons in the Solar System – and astronomers are on the look out for more. The largest four are huge, round worlds in their own right. About a dozen are tens of kilometres across, and the rest are just a few kilometres wide and irregular in shape.

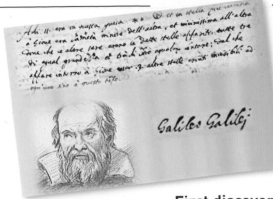

First discovery
Italy's Galileo Galilei was the first to see Jupiter's four largest moons, in 1610. They were the first moons to be found after Earth's Moon, and are known as the Galileans. A portrait of Galileo and his description of a 1610 observation is on the Juno spacecraft now at Jupiter.

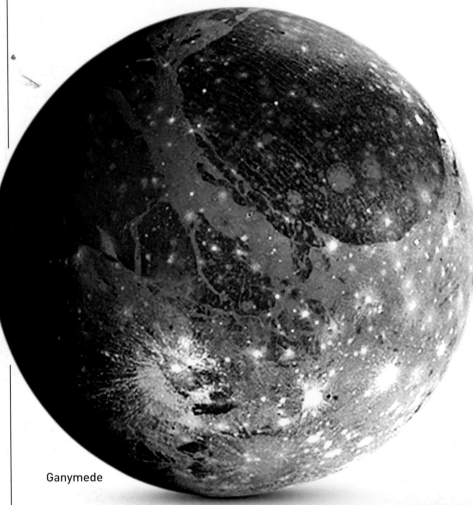

Ganymede

Callisto

The Galilean moons
Ganymede, Callisto, Io, and Europa are known as the Galilean moons. Ganymede is bigger than the planet Mercury and is the largest moon in the Solar System. It is an icy world – so too are Callisto and Europa. Io is the closest of the four to Jupiter, and is covered in volcanoes. Jupiter's gravity distorts Io's shape and generates internal heat that is released at the moon's surface.

Ganymede's icy surface
This huge ball of rock and ice has an icy crust of bright and dark areas – the bright regions are dominated by ice-tipped ridges and grooves, the darker ones by impact craters. Enki Catena is a chain of 13 craters that runs across a boundary between the two areas for 161 km (100 miles). It formed as fragments of an asteroid or comet crashed into Ganymede.

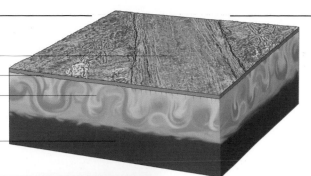

Ice-filled cracks called linea criss-cross Europa's surface.

Solid crust of ice

Warm ice layer

Liquid water ocean is about 100 km (62 miles) deep

Europa and water

The smallest and second most distant of the Galilean moons, Europa is squeezed and stretched by the gravitational pull of Jupiter on one side and Ganymede on the other. An ocean of liquid water beneath its frozen crust is one of the few places in the Solar System that could be hospitable to life. Ganymede also has an ocean, roughly 200 km (124 miles) below its surface, between layers of ice.

Triple eclipse

Jupiter's largest moons regularly travel across its face, but the sight of three at a time occurs only once or twice a decade. In this view by the Hubble Space Telescope in January 2015, Callisto, Europa, and Io are all passing between Jupiter and the Sun, casting shadows on the planet as they eclipse the Sun.

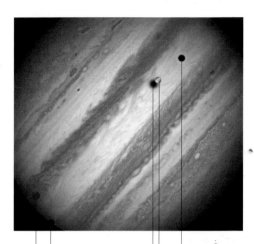

Callisto | Europa's shadow | Callisto's shadow | Io | Io's shadow

Volcanic Io

Io orbits Jupiter every 1.7 days at slightly more than the Earth–Moon distance. As the most volcanically active body in the Solar System, Io is covered by lava flows, and towering plumes of gas and rock stream from its volcanic craters.

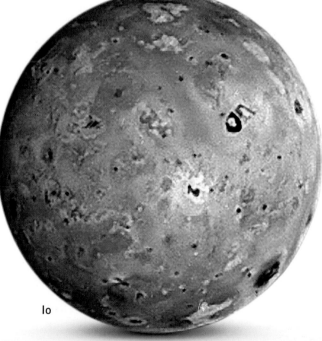

Io

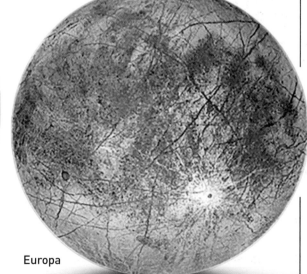

Europa

21st-century discoveries

More moons, all much smaller than the Galileans, were discovered between 1610 and 2000. Smaller still are those discovered since 2000. More than 40 have been detected by a team led by Scott S Sheppard at the Mauna Kea Observatory in Hawaii, USA (below). They average only 3 km (1.9 miles) across and orbit clockwise (the opposite direction to the large moons), which suggests they are captured asteroids.

Saturn

In ancient times, Saturn was the most distant planet known, and marked the edge of the Solar System. We now know this giant planet is the second largest, the sixth of eight orbiting the Sun, with a wide system of rings and a large family of moons. Cassini has been orbiting Saturn since 2004.

Early observations

In 1610, Italy's Galileo Galilei, using the newly invented telescope, was the first to see Saturn's ring system. He thought it was ear-like moons, one at either side of the planet. As telescopes improved, Dutch astronomer Christiaan Huygens (above) saw the rings' true nature in the 1650s and discovered the first of Saturn's moons, Titan, in 1655.

Early drawings of Saturn, from Christiaan Huygens' *Systema Saturnium* (1659)

Saturn's rings make an almost flat plane around the planet.

Ringed world

Like Jupiter, Saturn is mainly hydrogen and helium, gaseous in its outer layer but liquid inside. Not quite as wide as Jupiter, it has less than a third of its material, making Saturn the least dense of all the planets. It is also the least spherical: its low density and fast spin of less than 11 hours fling its equatorial regions out to form a bulging waist.

The Cassini Division looks empty from Earth but is full of ring material.

Banded atmosphere

The upper layer of Saturn's atmosphere forms bands with clouds and storms driven by winds of up to 1,800 kph (1,120 mph). Saturn's largest storms appear as white spots in its northern hemisphere every 20–30 years, at the onset of its summer.

Northern storm

The most intense storm ever seen on Saturn took shape in the northern hemisphere in 2011 (above). Within three months, it had spread right around the planet.

Saturn is 10 per cent wider at its equator than at its poles.

The central hurricane is about 50 times larger than any on Earth.

Polar regions

Saturn's atmospheric bands produce a hurricane-like vortex at each pole. The northern vortex (right) is surrounded by a six-sided jet stream that whips around at about 350 kph (217 mph). In this false-colour close-up, other storms appear as reddish or white ovals.

Earth

The view from Saturn

In this Cassini image, Earth appears as a tiny, distant, pale blue dot. Saturn is 9.5 times further from the Sun than Earth and takes 29.5 years to orbit the Sun. It receives about one per cent of the Sun's heat and light that reaches Earth, but creates more heat internally.

Saturn's rings

The most impressive rings of any planet encircle Saturn. They are made of millions of orbiting pieces prevented by Saturn's gravity from combining to form a single moon. The rings extend to many times Saturn's width but average only about 10 m (33 ft) deep, and small moons sweep the gaps in between.

Giovanni Cassini
Italian-born Cassini, the first director of the Paris Observatory, France, was one of the first to observe Saturn. In 1675, he spotted the gap dividing the A and B rings that now bears his name. He also discovered four moons: Iapetus, Rhea, Tethys, and Dione.

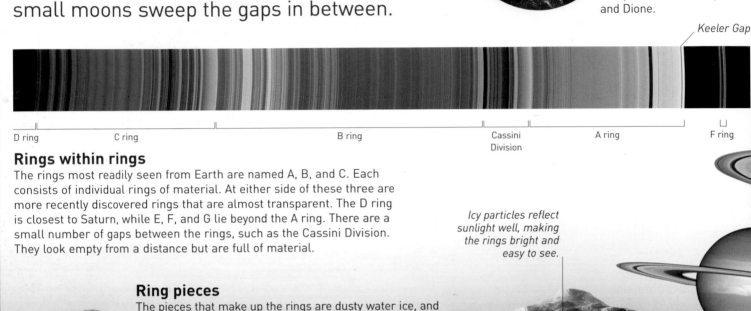

Keeler Gap

D ring   C ring   B ring   Cassini Division   A ring   F ring

Rings within rings

The rings most readily seen from Earth are named A, B, and C. Each consists of individual rings of material. At either side of these three are more recently discovered rings that are almost transparent. The D ring is closest to Saturn, while E, F, and G lie beyond the A ring. There are a small number of gaps between the rings, such as the Cassini Division. They look empty from a distance but are full of material.

Icy particles reflect sunlight well, making the rings bright and easy to see.

Ring pieces

The pieces that make up the rings are dusty water ice, and range in size from tiny grains to truck-sized boulders. Each follows its own circular orbit in a plane extending out from Saturn's equator. Their origin is uncertain. The pieces could be debris from a moon torn apart by Saturn's gravity, or from a moon destroyed in a collision with another body.

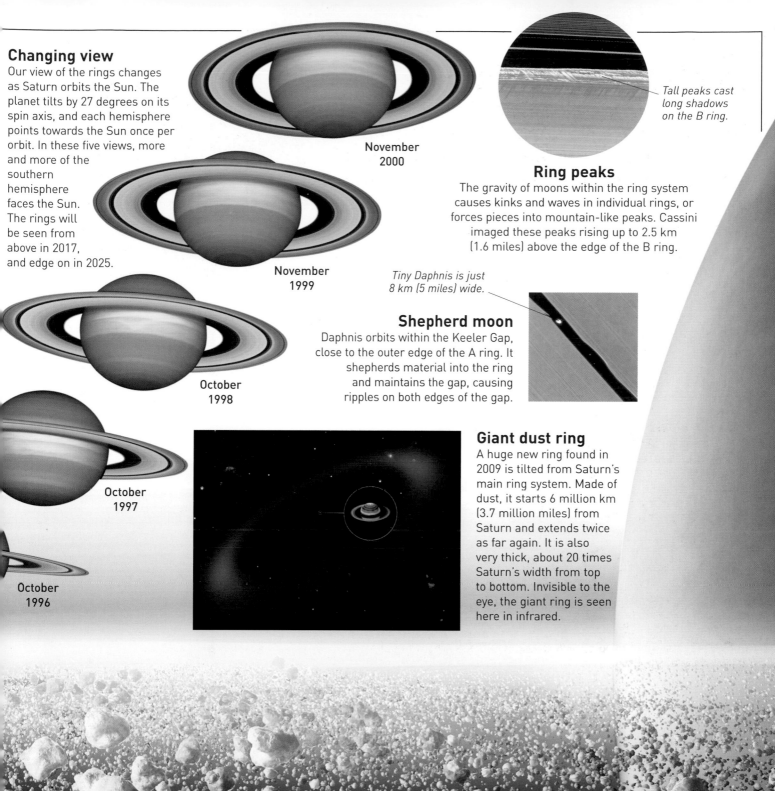

Changing view

Our view of the rings changes as Saturn orbits the Sun. The planet tilts by 27 degrees on its spin axis, and each hemisphere points towards the Sun once per orbit. In these five views, more and more of the southern hemisphere faces the Sun. The rings will be seen from above in 2017, and edge on in 2025.

November 2000

November 1999

October 1998

October 1997

October 1996

Tall peaks cast long shadows on the B ring.

Ring peaks

The gravity of moons within the ring system causes kinks and waves in individual rings, or forces pieces into mountain-like peaks. Cassini imaged these peaks rising up to 2.5 km (1.6 miles) above the edge of the B ring.

Tiny Daphnis is just 8 km (5 miles) wide.

Shepherd moon

Daphnis orbits within the Keeler Gap, close to the outer edge of the A ring. It shepherds material into the ring and maintains the gap, causing ripples on both edges of the gap.

Giant dust ring

A huge new ring found in 2009 is tilted from Saturn's main ring system. Made of dust, it starts 6 million km (3.7 million miles) from Saturn and extends twice as far again. It is also very thick, about 20 times Saturn's width from top to bottom. Invisible to the eye, the giant ring is seen here in infrared.

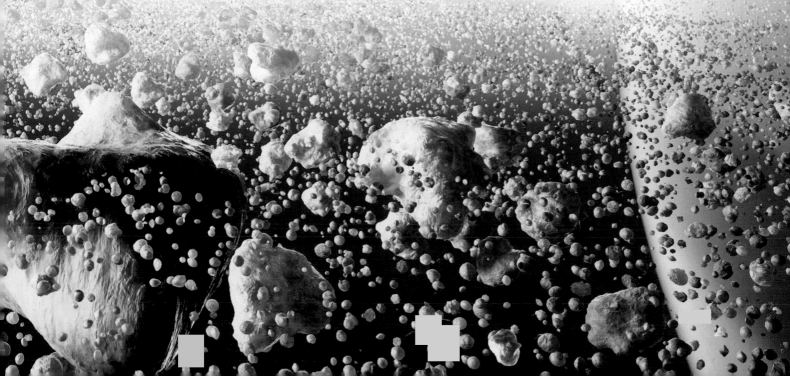

Saturn's moons

Saturn's 62 moons range from Titan, bigger than Mercury, to more than 30 city-sized icy rocks. The bigger moons orbit closest to Saturn – some within its rings – while the smallest orbit up to 25 million km (15.5 million miles) away. All are icy worlds, but Titan is the only Solar System moon with a substantial atmosphere and liquid seas and lakes.

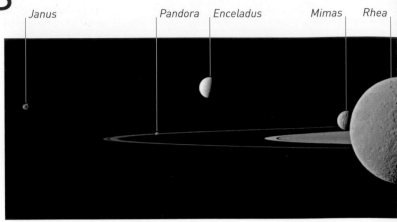

Janus   Pandora Enceladus   Mimas   Rhea

Little and large
Saturn's seven major moons are Titan, Rhea, Iapetus, Dione, Tethys, Enceladus, and Mimas. All seven are round and relatively close to Saturn. In this Cassini image, three of the major moons are seen along with smaller moons Janus and Pandora.

Hyperion
Some of Saturn's moons are irregular in shape. Hyperion, the largest of these, is 360 km (224 miles) long. This icy rock body looks sponge-like – hit by other objects in the past, its surface material was blasted away. It orbits Saturn in 21.3 days, rotating chaotically as it travels.

Explosive Enceladus
When Cassini flew by Enceladus, it saw that huge areas have been resurfaced and that it is geologically active. Jets of water ice and water vapour burst through four long fractures in its crust. The material that shoots through them supplies the E ring.

Ithaca Chasma is a canyon system that runs for 1,219 km (757 miles).

Herschel Crater is named after William Herschel, who discovered Mimas.

Long fractures cut through the icy surface.

Mimas
The innermost and smallest of Saturn's major moons orbits in the Cassini Division.

Enceladus
This bright moon orbits within the E ring, beyond the main rings. It is 512 km (318 miles) across.

Tethys
Like Mimas and Enceladus, Tethys is made of water-ice and rock. It shares its orbit with two small, irregular-shaped moons.

Dione
Long, white streaks on Dione's cratered surface are deep canyons in the icy crust. Dione orbits within the E ring, with two small, irregular moons.

Seas on Titan

Titan's three large seas are all close to the
north pole. Ligeia Mare (above) is the second
largest, and mainly methane. It is about
500 km (311 miles) at its widest point,
and up to 160 m (525 ft) deep.

Titan

The second-largest moon
in the Solar System is a cold
world of rock and ice about
5,150 km (3,200 miles) across,
hidden by a thick nitrogen
atmosphere. Cassini used
a radar mapper to see
through to the surface,
and released a probe
to land on it in 2005,
unveiling a world of
mountains, seas,
lakes, and dunes.

Huygens probe

Cassini's Huygens probe parachuted on to Titan's surface
and sent back data for about 90 minutes. The landing site
was flat and strewn with smooth pebbles and rocks.

*Engelier
Crater*

*Bright icy
material thrown out
by an impact crater*

Iapetus

Iapetus takes 79.3 days to orbit, at an average
distance of 3.6 million km (2.2 million miles) from Saturn.
One side is dark, the other, light, with craters on both –
Engelier, one of the largest, is 504 km (313 miles) wide.

Rhea

Rhea is Saturn's second-largest moon – 1,528 km (949 miles)
across – and the first of the moons that orbit Saturn beyond
the ring system, completing its orbit in 4.5 days. Rhea's icy
surface is heavily pockmarked by impact craters.

Where are they now?

All four Pioneer and Voyager craft continue to travel away from the Sun. The year of launch, and the year and distance when we last received a signal from Pioneer 10 and 11, are shown above. The Voyagers are still sending signals from beyond the planets.

Visiting the giants

The first spacecraft to venture beyond the Asteroid Belt and fly by the outer planets were Pioneer 10 and 11 and Voyager 1 and 2, which launched in the 1970s. In-depth study came later, with Galileo and Cassini orbiting Jupiter and Saturn respectively.

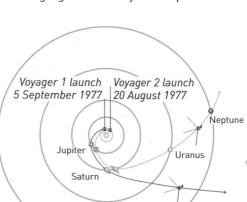

Voyager 1 launch 5 September 1977   Voyager 2 launch 20 August 1977

Neptune
Jupiter
Uranus
Saturn

The grand tour

The Voyager craft left Earth in 1977. Voyager 1 flew by Jupiter and Saturn, and was the first to leave the planetary part of the Solar System when it entered interstellar space in 2012. Voyager 2 flew past Jupiter and Saturn, and then past Uranus and Neptune. By harnessing the gravity of each planet, it swung from one to the next, using less fuel.

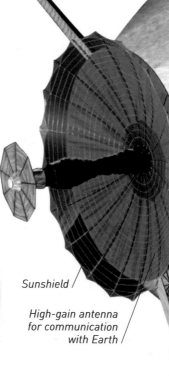

An 11-m (36-ft) boom carried Galileo's magnetometers.

A nuclear generator provided Galileo with power.

Sunshield

High-gain antenna for communication with Earth

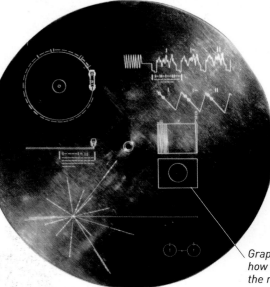

Graphics show how to play the record.

Message from Earth

Each Pioneer carries a metal plaque identifying where it comes from. The two Voyagers carry a gramophone record that locates Earth and includes many of its sounds and images. Information on the cover (above) is designed to enable any advanced civilization to play the record. A stylus (needle) for playing the record is also provided.

Galileo at Jupiter

In 1995–2003, Galileo (above) made the first long-term study of Jupiter in close-up, with 35 orbits of the planet and flybys of its Galilean moons. It also released a probe (right) into Jupiter's atmosphere. At the end of the mission, Galileo destroyed itself by plunging into the giant planet to avoid contaminating its moons.

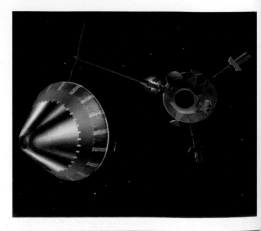

PIONEER 10
(1972–2003)
12.2 billion km (7.6 billion miles)

VOYAGER 2
(1977– ongoing)
16.7 billion km (10.4 billion miles)

VOYAGER 1
(1977– ongoing)
20.3 billion km (12.6 billion miles)

Cassini's main radio dish is 4 m (13 ft) across.

Cassini's mission to Saturn

The size of a bus, Cassini is the largest and most complex craft sent to a planet. In 2004, it started orbiting Saturn (below). On an early flyby of the moon Titan, it released the Huygens probe on to Titan's surface. Cassini's mission ends in 2017, after 20 years of service – 7 travelling to Saturn and 13 in orbit.

Cassini recorded ring particles and temperature as it orbited Saturn.

Holes in the probe's cover let in atmospheric gas for analysis.

Huygens' descent to Titan

On release by Cassini, Huygens' main parachute opened to slow its descent to Titan. As its heatshield fell away, the probe began to test the atmosphere and image the moon's surface. It transmitted data as it fell and for about 1.5 hours after landing.

Shield of heat-resistant tiles protected Huygens on its descent

The way ahead

Juno began orbiting Jupiter in 2016, measuring its magnetic and gravitational fields. Launching in the 2020s, Europa (above) will study Jupiter's moon Europa, while JUICE (JUpiter ICy moons Explorer) will study Ganymede, Callisto, and Europa.

The outer Solar System

The first object discovered beyond Saturn was Uranus, in 1781. Today, we know this dark and cold outer region of the Solar System contains billions of objects – not only the planets Uranus and Neptune, but dwarf planets and a ring of Kuiper Belt objects, too. Beyond, surrounding all of this, is the Oort Cloud, home to comet nuclei.

Neptune

The Oort Cloud consists of billions of comet nuclei on individual orbits around the Sun.

The Sun and the planets are in the centre of the Oort Cloud.

Beyond Saturn

The eight planets orbit the Sun in roughly the same plane, making a disc-shaped system. Beyond is the Kuiper Belt, a flattened belt thought to contain hundreds of thousands of ice-and-rock bodies. Many of these, such as Pluto, follow orbits that take them out of the planetary plane. The Belt and the Oort Cloud are made up of material left over after the planets formed at the dawn of the Solar System.

The Oort Cloud

The vast Oort Cloud is a reservoir of comet nuclei that follow elongated paths around the Sun in all directions. As a group, they form the spherical Oort Cloud, which surrounds the rest of the Solar System. The outer edge of the cloud is about 100,000 times more distant from the Sun than Earth is, and stretches halfway to the nearest stars.

1992 QB1 on 27 September 1992 at 02:35 hours

Kuiper Belt objects

More than 1,000 ice-and-rock bodies are known in the Kuiper Belt. Most are classed as Kuiper Belt objects (KBOs), some are comet nuclei, and the largest are dwarf planets such as Pluto. When the first KBO beyond Pluto was discovered in 1992, it was the most distant Solar System body known. Labelled 1992 QB1, it was first seen by astronomers at the Mauna Kea Observatory in Hawaii, who tracked its progress and calculated its orbit.

1992 QB1 on 27 September at 06:42

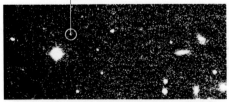

Planet discovery

William Herschel (left) was a German-born astronomer working in England. When he found Uranus in 1781, no one knew planets existed beyond Saturn. Scientists then realized that something was pulling on Uranus as it orbited the Sun, and began a hunt for a more distant planet. In 1846, German astronomer Johann Galle discovered Neptune.

1992 QB1 on 28 September at 06:58

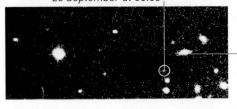

Background stars remain fixed while 1992 QB1 moves against their backdrop.

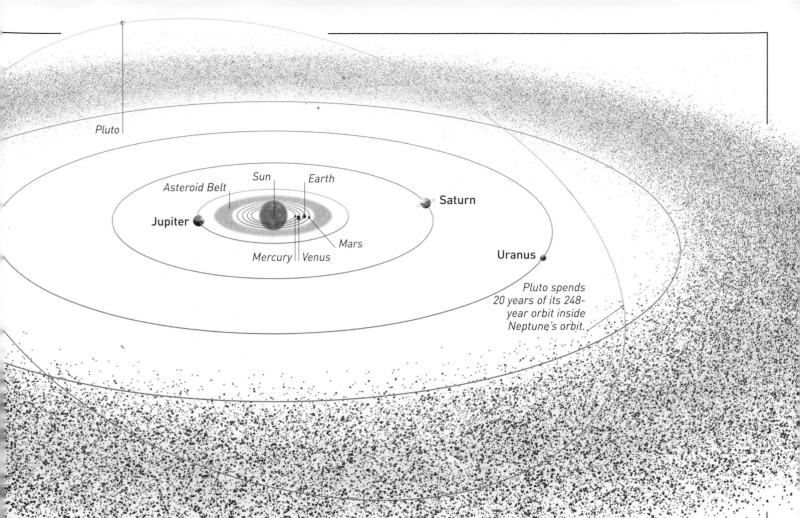

Pluto

Asteroid Belt

Sun

Earth

Jupiter

Mercury | Venus

Mars

Saturn

Uranus

Pluto spends 20 years of its 248-year orbit inside Neptune's orbit.

Makemake is about 1,430 km (889 miles) across and has a moon.

Dwarf planets
The first Solar System body known beyond Neptune was Pluto. Classed as a planet since its discovery in 1930, Pluto differed from the others in size, make up, and orbit, and has a huge moon, Charon. Pluto is now classed as one of four dwarf planets in the Kuiper Belt. The others are Eris, Haumea, and Makemake (above).

Charon

Pluto

What's in a name?
Some astronomers refer to the Kuiper Belt and Oort Cloud as the Edgeworth-Kuiper Belt and Öpik-Oort Cloud. Kenneth Edgeworth predicted a belt in 1943 – in 1951, Gerard Kuiper said it no longer existed. Ernst Öpik suggested the cloud of comets in 1932. Jan Oort (right) revived the idea in the 1950s.

Artist's impression of the view from Sedna

Most distant
Some objects follow orbits out beyond the Kuiper Belt. Like Eris, they could be dwarf planets. They might also be the first inner Oort Cloud objects to be found. Sedna (left) is 90 times the Earth–Sun distance. The furthest known is V774104, at more than 100 times the Earth–Sun distance.

Uranus

Barely visible to the naked eye, and twice as far from the Sun as its inner neighbour Saturn, Uranus is a freezing cold world. Only Voyager 2 has visited this ice giant, flying by in 1986. Spinning on its side as it makes its 84-year orbit of the Sun, Uranus has faint, dark rings and 27 moons, which are much smaller than Earth's Moon.

Some parts of Uranus's atmosphere are –224°C (–371°F), colder than any other planet.

Pale blue ball

Uranus is four times the width of Earth and is the Solar System's third-largest planet. It is made mainly of water, ammonia, and methane. They form a slushy liquid mantle around a core of molten iron and magma. The outer layer is an atmosphere of hydrogen and helium. Methane gas in the atmosphere absorbs the red wavelengths of incoming sunlight and gives Uranus its blue colouring.

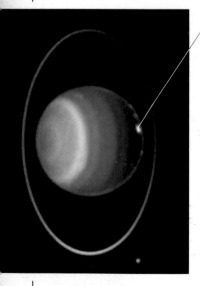

Clouds appear white, and pink around the planet's edge represents a high-altitude haze.

Bright clouds

When Voyager 2 flew by, the south pole faced the Sun and the planet looked featureless. But when the equator faces the Sun, new parts of Uranus are warmed up and it becomes a dynamic world. Infrared imaging (left) highlights its banded structure and bright clouds.

Tilted planet

Uranus's spin axis is tilted at almost right angles to its orbit, so it rolls along its path like a ball. Its rings and moons seem to orbit it from top to toe. During the planet's orbit, each pole points once to the Sun, receiving 42 years of sunlight, followed by 42 years of darkness.

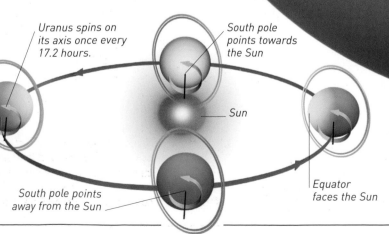

Uranus spins on its axis once every 17.2 hours.

South pole points towards the Sun

Sun

South pole points away from the Sun

Equator faces the Sun

The inner rings consist of dust and dark, rocky material from a few centimetres to several metres across.

Inner and outer rings

Uranus has 13 narrow rings made of dust and dark, rocky material – 11 form an inner ring system, and two are more distant. In this false-colour image of the inner rings, the bright white one is Epsilon, the furthest from Uranus. Closer and thinner rings, just a few kilometres across, are pale green, pale blue, and cream.

Hubble discoveries

The first of Uranus's rings were discovered in 1977, when astronomers saw them block the light of a distant star. Others were detected by Voyager 2 in 1986, and Hubble found two new outer rings – Nu and Mu – in 2003. This image combines Hubble's grainy images of Nu and Mu with an older view of Uranus in colour.

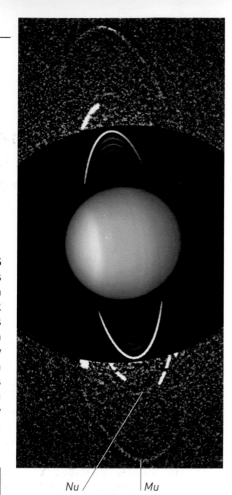

Nu *Mu*

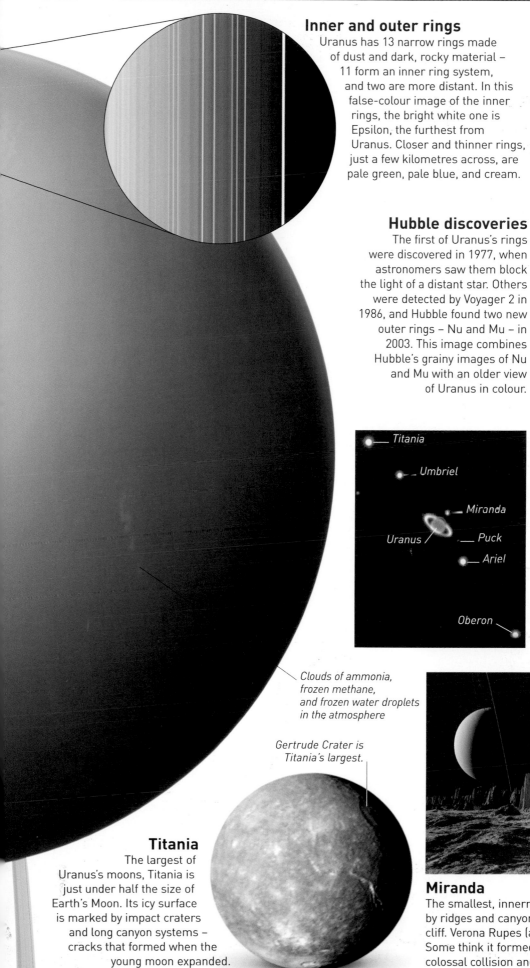

Titania

Umbriel

Miranda

Uranus *Puck*

Ariel

Oberon

Moons

Only Uranus's five largest moons are roughly spherical. From largest to smallest, they are Titania, Oberon, Umbriel, Ariel, and Miranda. The other 22 are smaller and irregular in shape. The closest orbit Uranus in just hours, while the five largest take between one and 13 days, and the furthest take years. Tiny Puck, just 162 km (101 miles) across, orbits in 18 hours.

Clouds of ammonia, frozen methane, and frozen water droplets in the atmosphere

Gertrude Crater is Titania's largest.

Titania

The largest of Uranus's moons, Titania is just under half the size of Earth's Moon. Its icy surface is marked by impact craters and long canyon systems – cracks that formed when the young moon expanded.

Miranda

The smallest, innermost of the major moons is criss-crossed by ridges and canyons, and has the Solar System's tallest cliff. Verona Rupes (above) is almost 10 km (6 miles) high. Some think it formed when the moon split apart after a colossal collision and came together again.

Neptune

The most distant planet in the Solar System, Neptune is 30 times further from the Sun than Earth. It was also the last to be discovered, in 1846, and the last to be visited by a spacecraft, Voyager 2 in 1989. Neptune has the longest orbital path around the Sun, completing just one 165-year orbit since its discovery.

Neptune's rings are so thin and sparse that they appear transparent to the eye.

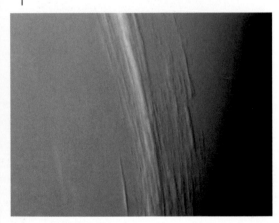

Deep blue world

Neptune is an ice giant like Uranus and is made of mainly water, ammonia, and methane. Beneath its atmosphere is a slushy layer of liquid and ice, and deep inside is a core of rock and iron. It is a little smaller than Uranus but more massive, due to its thinner atmosphere and deeper liquid layer. Its spin axis is tilted at a similar angle to Earth's, and it experiences seasons – each season lasts about 40 years.

Atmosphere

Neptune's atmosphere is mainly hydrogen and helium, with water, ammonia, and methane increasing with depth. Fast winds whip around the planet at up to 2,100 kph (1,300 mph), ten times faster than hurricanes on Earth, and clouds sporadically appear.

Methane gas in the atmosphere absorbs red wavelengths in sunlight, giving Neptune its blue colour.

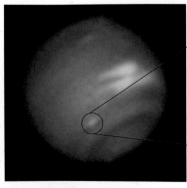

Neptune in May 2016

Dark spot with white clouds above it

Dark spots

Huge storms on Neptune appear as dark spots. The Great Dark Spot – seen by Voyager 2 in 1989 – was almost as big as Earth, but disappeared a few years later. A spot the size of the United States appeared in May 2016 (above, captured by the Hubble Space Telescope).

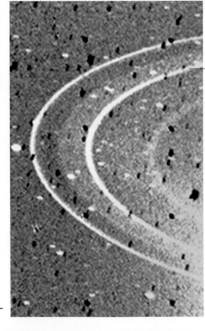

Le Verrier ring and the fainter, broad Lassell ring extending beyond it

Rings

Neptune's five dark, dusty rings are difficult to see except when lit by the Sun from behind. Their presence had been predicted, but it was Voyager 2's flyby that confirmed the rings' existence. It also found the moon Despina, shepherding particles in the Le Verrier ring, and Galatea, which shepherds particles in the brighter outer Adams ring.

Triton

About three-quarters the size of our Moon, Triton is a ball of rock and ice and orbits Neptune in less than six days. Its surface temperature of –235°C (–391°F) is one of the coldest in the Solar System. Pink methane ice is found near its south pole (right).

Plumes of nitrogen gas and dust reach 8 km (5 miles) above Triton's surface.

Ice volcanism

Triton is ice-cold but volcanically active. Voyager 2's images of the surface revealed smooth plains of volcanic ice, and mounds and pits formed by icy lava. Beneath is a mantle of ice that erupts through surface cracks. Heated by the Sun, the nitrogen ice turns to gas, and plumes of gas and dust jet up high above the surface.

Adams ring is Neptune's outermost ring

Moons

Most of Neptune's 14 moons are small and irregular, but Triton is spherical and 2,707 km (1,682 miles) wide. It was discovered in 1846, just a couple of weeks after Neptune, but no others were found until Nereid, in 1949. The latest and smallest, S/2004 N1, was discovered in 2013 by the Hubble Space Telescope.

Methane ice clouds in the atmosphere change their appearance in hours.

S/2004 N1 · Psamathe · Laomedeia · Sao · Neso · Halimede · Naiad · Thalassa · Despina · Galatea · Larissa · Nereid · Proteus · Triton

The dwarf planets

Four dwarf planets orbit in the Kuiper Belt. Pluto was discovered in a search for Planet X – once believed to exist beyond Neptune – and the others, in a hunt for Kuiper Belt objects. All four are small, cold worlds of ice and rock. New Horizons revealed more detail when it flew by Pluto in 2015.

Seeing Pluto close-up

New Horizons' images of Pluto showed mountains as high as Earth's Rockies, and the Solar System's largest glacier – the crater-free, central region called Sputnik Planitia. This 1,000-km- (620-mile-) wide ice plain is constantly renewed as older nitrogen ice is replaced by newer material rising from underneath.

Clyde Tombaugh with his own reflecting telescope in 1928

Discovery

American astronomer Clyde Tombaugh joined the Lowell Observatory in Arizona, USA, in 1929. In a search for Planet X, he photographed the same region of sky on different nights and compared the images to identify objects that had moved. In February 1930, he found the object later named Pluto.

Mountains capped with methane snow and ice, in the dark Cthulu Regio

Norgay Montes – ice mountains up to 3.5 km (2.2 miles) high – captured by New Horizons in July 2015

Frozen landscape

Pluto has snow-capped mountain ranges, ice plains, vast canyons, and craters formed by impacts with other Kuiper Belt bodies. During Pluto's 248-year orbit, its thin nitrogen atmosphere freezes to ice and snow, and then evaporates again with the changing seasons.

Sputnik Planitia

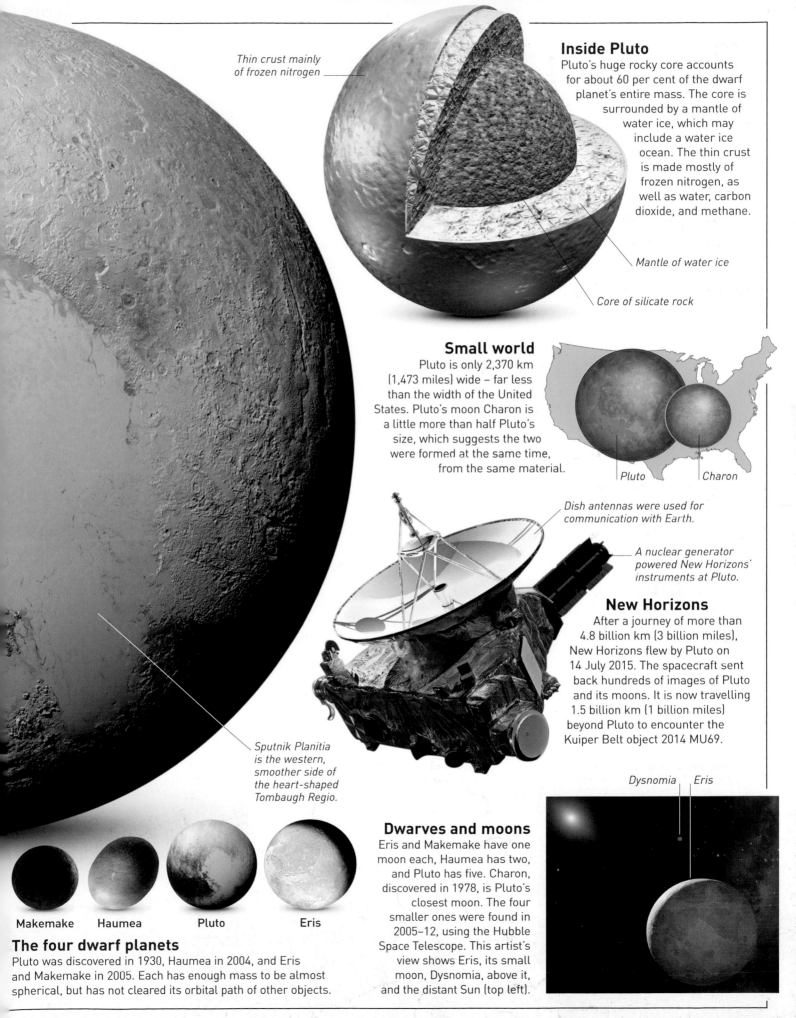

Thin crust mainly of frozen nitrogen

Inside Pluto

Pluto's huge rocky core accounts for about 60 per cent of the dwarf planet's entire mass. The core is surrounded by a mantle of water ice, which may include a water ice ocean. The thin crust is made mostly of frozen nitrogen, as well as water, carbon dioxide, and methane.

Mantle of water ice

Core of silicate rock

Small world

Pluto is only 2,370 km (1,473 miles) wide – far less than the width of the United States. Pluto's moon Charon is a little more than half Pluto's size, which suggests the two were formed at the same time, from the same material.

Pluto   Charon

Dish antennas were used for communication with Earth.

A nuclear generator powered New Horizons' instruments at Pluto.

New Horizons

After a journey of more than 4.8 billion km (3 billion miles), New Horizons flew by Pluto on 14 July 2015. The spacecraft sent back hundreds of images of Pluto and its moons. It is now travelling 1.5 billion km (1 billion miles) beyond Pluto to encounter the Kuiper Belt object 2014 MU69.

Sputnik Planitia is the western, smoother side of the heart-shaped Tombaugh Regio.

Dysnomia   Eris

Makemake   Haumea   Pluto   Eris

The four dwarf planets

Pluto was discovered in 1930, Haumea in 2004, and Eris and Makemake in 2005. Each has enough mass to be almost spherical, but has not cleared its orbital path of other objects.

Dwarves and moons

Eris and Makemake have one moon each, Haumea has two, and Pluto has five. Charon, discovered in 1978, is Pluto's closest moon. The four smaller ones were found in 2005–12, using the Hubble Space Telescope. This artist's view shows Eris, its small moon, Dysnomia, above it, and the distant Sun (top left).

Comets

Visitors from the edge of the Solar System, comets are primitive material left over from when the planets formed. These city-sized dirty snowballs, each on its own orbit around the Sun, are mostly too distant to be seen. Every so often, when a star's gravity pushes one much closer to the Sun, it becomes a long-tailed comet.

This illustration of a comet nucleus reveals the snow, ice, and rock dust held together loosely by gravity.

Dirty snowball nucleus

In the 1950s, American Fred Whipple suggested that the long-tailed comets we see in our skies are produced by a huge, dirty snowball nucleus. In 1986, the Giotto spacecraft imaged the nucleus of Comet Halley. It was a mix of two-thirds snow and ice, one-third rock-dust, 15.3 km (9.5 miles) in length.

Towards the Sun

A comet becomes big and bright only as it orbits closer to the Sun. The Sun's heat turns its snow and ice to gas, which jets out from the nucleus, taking dust with it. A huge coma (head) forms around the nucleus, and two tails – one of gas, the other, dust. They shrink again as it moves away.

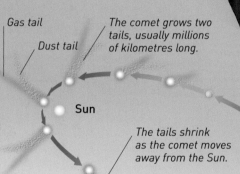

Gas tail

Dust tail

The comet grows two tails, usually millions of kilometres long.

Sun

The tails shrink as the comet moves away from the Sun.

The surface loses a layer 1 m (3.3 ft) thick on each close orbit of the Sun.

Great comets

About 10 comets a century are bright enough to be easily seen in the night sky and are classed as great comets. Historically, they were named for the year they appeared. Today, they take their discoverer's name. The brightest one so far this century was Comet McNaught in 2007 (above).

Periodic comets

Some comets follow orbits that bring them back time and again to Earth's sky, and are called periodic comets. Some take many thousands of years to return. Others return more frequently. Halley reappears about every 76 years. It shines above the nativity scene in *The Adoration of the Magi* (left), painted by Italian artist Giotto di Bondone in about 1305.

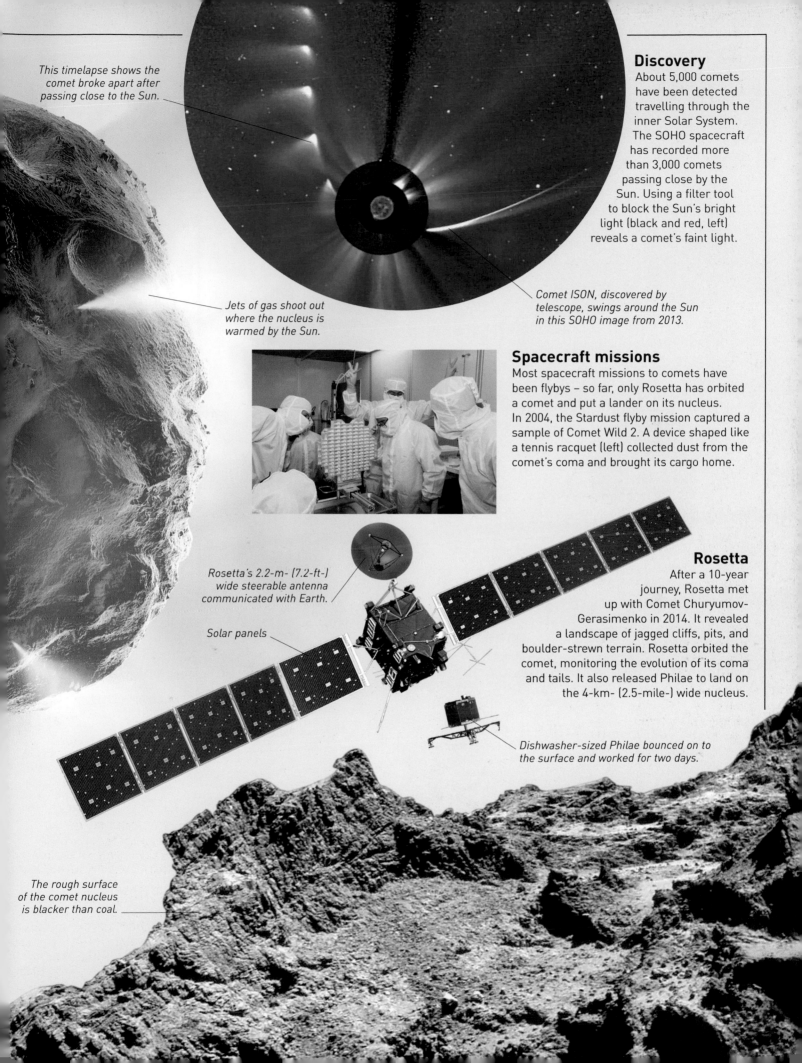

This timelapse shows the comet broke apart after passing close to the Sun.

Jets of gas shoot out where the nucleus is warmed by the Sun.

Discovery

About 5,000 comets have been detected travelling through the inner Solar System. The SOHO spacecraft has recorded more than 3,000 comets passing close by the Sun. Using a filter tool to block the Sun's bright light (black and red, left) reveals a comet's faint light.

Comet ISON, discovered by telescope, swings around the Sun in this SOHO image from 2013.

Spacecraft missions

Most spacecraft missions to comets have been flybys – so far, only Rosetta has orbited a comet and put a lander on its nucleus. In 2004, the Stardust flyby mission captured a sample of Comet Wild 2. A device shaped like a tennis racquet (left) collected dust from the comet's coma and brought its cargo home.

Rosetta's 2.2-m- (7.2-ft-) wide steerable antenna communicated with Earth.

Solar panels

Rosetta

After a 10-year journey, Rosetta met up with Comet Churyumov-Gerasimenko in 2014. It revealed a landscape of jagged cliffs, pits, and boulder-strewn terrain. Rosetta orbited the comet, monitoring the evolution of its coma and tails. It also released Philae to land on the 4-km- (2.5-mile-) wide nucleus.

Dishwasher-sized Philae bounced on to the surface and worked for two days.

The rough surface of the comet nucleus is blacker than coal.

Exoplanets

We know of more than 3,500 planets orbiting other stars, with more found each month. Ranging from more massive than Jupiter to near Earth-like, exoplanets orbit a huge variety of distances around different stars. Some orbit as part of a family, but few multi-systems resemble our Solar System.

First discovery

In the 1980s, discs of material from which planets can form were identified around distant stars. Some of the gas and dust around Fomalhaut (above) formed the exoplanet Fomalhaut b, now tracked as it orbits along the inner edge of the ring (right).

2004 | 2006

Solar panels provide power for Kepler and its equipment.

Spacecraft search

More than 2,300 exoplanets have been discovered using the Kepler spacecraft, with a similar number waiting to be confirmed. In 2009–13, Kepler studied an area of sky, monitoring 100,000 stars for Earth-like planets, and it has now widened its search. It uses a photometer, which records the stars' brightness. Subtle dips in brightness indicate an exoplanet in front of its star.

A radiator keeps instruments inside Kepler cool.

Far distant worlds

The exoplanets so far discovered orbit about 2,600 stars. The stars are too bright for the exoplanets to be seen directly, but many are revealed by their gravity causing their star to wobble. Others show up as they cross in front of a star and dim its light. Most exoplanets orbit a single star, but about 20 orbit two. Kepler-16b (above) is about the size of Saturn and orbits two stars in 229 days.

Thrusters keep the spacecraft in position.

A high-gain antenna transmits data to Earth.

Planets around other suns

The first exoplanet found around a Sun-like star was 51 Pegasi b, in October 1995. It is a hot-Jupiter type of exoplanet – massive like Jupiter and hot due to its closeness to its star. It travels around the star 51 Pegasi in just four days. Exoplanets take the name of their star, or the name of the spacecraft or project that discovered them, followed by a letter indicating the order of discovery (starting with b).

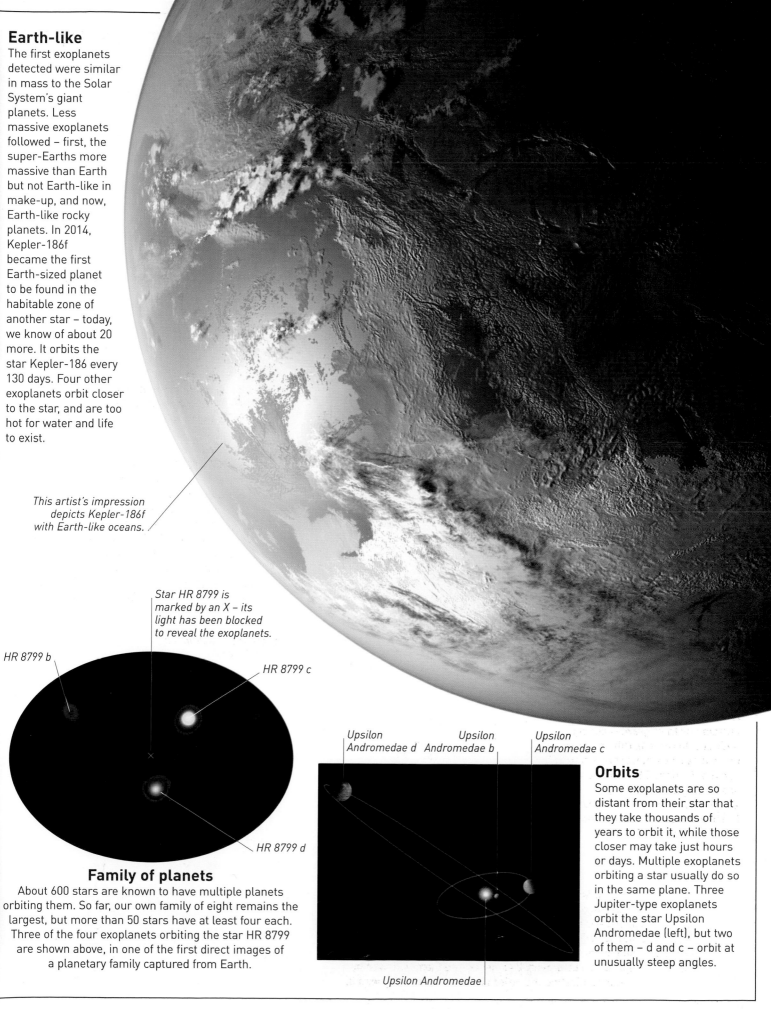

Earth-like

The first exoplanets detected were similar in mass to the Solar System's giant planets. Less massive exoplanets followed – first, the super-Earths more massive than Earth but not Earth-like in make-up, and now, Earth-like rocky planets. In 2014, Kepler-186f became the first Earth-sized planet to be found in the habitable zone of another star – today, we know of about 20 more. It orbits the star Kepler-186 every 130 days. Four other exoplanets orbit closer to the star, and are too hot for water and life to exist.

This artist's impression depicts Kepler-186f with Earth-like oceans.

Star HR 8799 is marked by an X – its light has been blocked to reveal the exoplanets.

HR 8799 b

HR 8799 c

HR 8799 d

Family of planets

About 600 stars are known to have multiple planets orbiting them. So far, our own family of eight remains the largest, but more than 50 stars have at least four each. Three of the four exoplanets orbiting the star HR 8799 are shown above, in one of the first direct images of a planetary family captured from Earth.

Upsilon Andromedae d *Upsilon Andromedae b* *Upsilon Andromedae c*

Upsilon Andromedae

Orbits

Some exoplanets are so distant from their star that they take thousands of years to orbit it, while those closer may take just hours or days. Multiple exoplanets orbiting a star usually do so in the same plane. Three Jupiter-type exoplanets orbit the star Upsilon Andromedae (left), but two of them – d and c – orbit at unusually steep angles.

Did you know?

Fascinating Facts

More than 2,000 years ago, Chinese astronomers recorded long-tailed comets sweeping across Earth's night sky. They called them broom stars, or hairy stars.

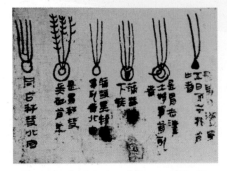

Comets depicted on Mawangdui silk texts (c.300 BCE) from tombs in Hunan, China

Some 19th-century astronomers thought a planet they named Vulcan existed between the Sun and Mercury. Vulcan only exists in the science-fiction TV and film series *Star Trek*, as a planet orbiting a distant star.

A new mineral – armalcolite – was found on the Moon in July 1969. It takes its name from the three Apollo 11 astronauts, Neil Armstrong, Buzz Aldrin, and Michael Collins.

Personal objects left by astronauts on the Moon include two golf balls left behind by Apollo 14's Alan Shepard in 1971, and a family photo placed there in 1972 by Charles Duke, the Apollo 16 astronaut and tenth man to walk on the Moon.

On 5 August 2013, one year after Curiosity's arrival on Mars, the rover sang "Happy Birthday" to itself. Technicians on Earth had programmed one of its soil-analysis instruments to vibrate and play the song.

Around 200 of the meteorites found on Earth are rocks from Mars. They were blasted off the Red Planet when asteroids crashed into it, and then followed orbits that brought them to Earth.

In 2014, astronomers discovered that the asteroid Chariklo has two dense, narrow rings – the first found other than the rings around the four giant planets. Chariklo is only 250 km (155 miles) wide and orbits the Sun between Saturn and Uranus.

Three toy LEGO® figures 3.8 cm (1.5 in) tall are on board the Juno spacecraft at Jupiter – the Roman god Jupiter, his wife Juno, and the astronomer Galileo Galilei.

The planet Uranus was discovered in 1781, but it may have been seen by the Greek astronomer Hipparchus in 128 BCE. He thought it was a star, as did England's Astronomer Royal, John Flamsteed, when he saw it in 1712.

The bunny on Mars

The Greek astronomer Hipparchus observing the stars

If you travelled to Neptune – the most distant planet in the Solar System – at 100 kph (62 mph), it would take you 500 years to get there.

A 1991 US postage stamp showing Pluto has travelled further than any other stamp. It is aboard the New Horizons spacecraft, which flew by Pluto in July 2015.

Historically, comets were seen as bad omens. When Halley's Comet passed by in 1910, Earth moved through its tail. Fearing it was poisonous, some people bought gas masks and anti-comet pills.

Images taken by the Opportunity rover after landing on Mars show what looks like a rabbit. Scientists concluded it was a piece of an airbag used to soften Opportunity's landing, or something similar.

The Opportunity rover, working on Mars since January 2004

Questions and Answers

Q Who names newly discovered planets and moons?

A Since the 1920s, the IAU (International Astronomical Union) has overseen the naming of space objects and their features, such as craters, mountains, and maria. Anyone can suggest a name to the IAU, but there are guidelines for each type of object. Dwarf planets, for example, are given the name of a god related to creation. Moons of Jupiter are named after the lovers and descendants of Jupiter, or his Greek equivalent, Zeus.

Q Who was the last astronaut on the Moon?

A The last mission to take astronauts to the Moon was Apollo 17 in December 1972. Two of its crew, Harrison Schmitt and Eugene Cernan, made three excursions together on to the Moon's surface. On the final trip, Cernan led the way out of the landing craft, making Schmitt the last man to step on to the Moon. A little over seven hours later, Schmitt was the first back into the craft, making Cernan the last to step off the Moon.

Harrison Schmitt collecting samples with a lunar rake

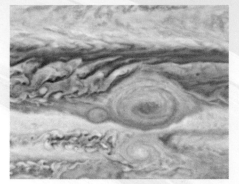

Jupiter's Great Red Spot (centre), with Red Spot Jr (below) and Baby Spot (left), June 2008

Q Does Jupiter have more than one red spot?

A The Great Red Spot is the huge, long-running storm that rages in Jupiter's southern hemisphere. Less well known are its short-lived, smaller companions. Red Spot Jr first appeared in early 2006 and moved across Jupiter's face below the Great Red Spot. A smaller spot – Baby Red Spot – appeared in 2008, but could not pass unscathed and was quickly consumed by the Great Red Spot.

Q Has Earth always spun round in 24 hours?

A Earth spun about four times faster when it was younger. Friction produced by both the Moon and Sun's tidal effect on Earth is slowing our planet's spin and moving the Moon away. Currently the Moon moves away by about 3.8 cm (1.5 in) each year. When dinosaurs roamed, Earth's spin was about 21 hours long.

Q Which is the nearest exoplanet to Earth?

A Proxima Centauri b orbits the closest star to the Sun, Proxima Centauri, about a quarter of a million times further from the Sun than Earth. A little more massive than Earth, the exoplanet orbits its star every 11 days, within its habitable zone.

Q How long does it take to send a message to a spacecraft?

A The further a spacecraft travels, the longer it takes to communicate with Earth. Messages to and from Curiosity on Mars take 14 minutes each way. Voyager 1's signals take 19 hours to reach us from interstellar space.

Q What is the largest telescope on Earth?

A The Gran Telescopio Canarias on La Palma in the Canaries, Spain, has the largest mirror, 10.4 m (34.1 ft) across. The European Extremely Large Telescope being built in Chile will have a 39.3-m- (129-ft-) wide mirror.

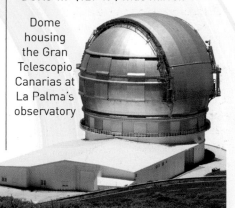

Dome housing the Gran Telescopio Canarias at La Palma's observatory

Record Breakers

LONGEST SPIN
Venus has the longest spin of any planet, rotating once in 243 days. Massive Jupiter has the shortest, spinning round once in 9.9 hours. Jupiter makes 589 spins during the time it takes Venus to make one.

SLOWEST ORBITER
The more distant a planet is from the Sun, the more slowly it orbits. Neptune is the furthest and slowest of the planets. It travels along its orbit at 5.4 km/s (3.4 miles/s). Closest to the Sun, Mercury is the fastest, speeding along at 47.4 km/s (29.5 miles/s).

GREATEST GRAVITATIONAL PULL
The more massive a planet, the greater its gravitational pull. Jupiter's is the greatest, at 2.4 times Earth's. If a rocket could launch from Jupiter, it would need to travel at 59.5 km/s (36.9 miles/s) to escape the planet's gravity.

Solar System facts

The eight planets form two groups. Smallest and closest to the Sun are the rocky planets Mercury, Venus, Earth, Mars, and just three moons. Further out and much colder, Jupiter, Saturn, Uranus, and Neptune have faster spins, longer orbits, and numerous moons.

SUN DATA

Diameter	1,393,684 km (865,374 miles)
Mass (Earth = 1)	333,000
Energy output	385 million billion gigawatts
Surface temperature	5,500°C (10,000°F)
Core temperature	15 million °C (27 million °F)
Average distance from Earth	149.6 million km (92.9 million miles)
Rotation period	25 days at the equator

PLANETARY DATA

	Mercury	Venus	Earth	Mars	Jupiter	Saturn	Uranus	Neptune
Diameter in km (miles)	4,879 (3,032)	12,104 (7,521)	12,756 (7,926)	6,792 (4,220)	142,984 (88,846)	120,536 (74,898)	51,118 (31,763)	49,528 (30,775)
Mass (Earth = 1)	0.06	0.82	1	0.11	317.83	95.16	14.54	17.15
Gravity (Earth = 1)	0.38	0.91	1	0.38	2.36	0.92	0.89	1.12
Rotation period in hours	1,407.6	5,832.5	23.9	24.6	9.9	10.7	17.2	16.1
Solar day (sunrise to sunrise) in hours	4,222.6	2,802	24	24.7	9.9	10.7	17.2	16.1
Average temperature	167°C (333°F)	464°C (867°F)	15°C (59°F)	−63°C (−81°F)	−108°C (−162°F)	−139°C (−218°F)	−197°C (−323°F)	−201°C (−328°F)
Closest distance to the Sun in km (miles)	46.0 million (28.6 million)	107.5 million (66.8 million)	147.1 million (91.4 million)	206.6 million (128.4 million)	740.5 million (460.1 million)	1,352.6 million (840.5 million)	2,741.3 million (1,703.4 million)	4,444.5 million (2,761.7 million)
Furthest distance from the Sun in km (miles)	69.8 million (43.4 million)	108.9 million (67.7 million)	152.1 million (94.5 million)	249.2 million (154.8 million)	816.6 million (507.4 million)	1,514.5 million (941.1 million)	3,003.6 million (1,866.4 million)	4,545.7 million (2,824.6 million)
Orbital period in days	87.96	224.70	365.24	686.97	4,330.59 (11.86 years)	10,746.94 (29.46 years)	30,588.74 (84.01 years)	59,799.9 (164.79 years)
Orbital speed in km/s (miles/s)	47.4 (29.5)	35.0 (21.7)	29.8 (18.5)	24.1 (14.9)	13.1 (8.1)	9.7 (6.0)	6.8 (4.2)	5.4 (3.4)
Number of moons	0	0	1	2	67	62	27	14

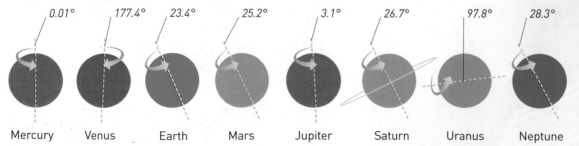

0.01°	177.4°	23.4°	25.2°	3.1°	26.7°	97.8°	28.3°
Mercury	Venus	Earth	Mars	Jupiter	Saturn	Uranus	Neptune

Spin and tilt

All eight planets spin around an axis (shown as a dotted line), tilted at an angle. Mercury is the most upright. Venus tilts so far that it spins in the opposite direction to the others. The time taken for one spin (rotation period) is shown in the table above – Earth takes 23.9 hours for one spin, but the four giant planets take less.

How far?

The vast distances between the Sun and the planets are described using the astronomical unit (au). One au is based on Earth's average distance from the Sun.

Key

au: astronomical unit
1 au = 149.6 million km (92.9 million miles)

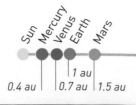

Sun · Mercury · Venus · Earth · Mars · Jupiter · Saturn

0.4 au | 0.7 au | 1 au | 1.5 au | 5.2 au | 9.5 au

WHAT DO YOU WEIGH?

Location	% of weight on Earth
Mercury	37.7
Venus	90.8
Earth	100
Moon	16.5
Mars	38.1
Jupiter	236
Saturn	92
Uranus	89
Neptune	112

Apollo 16's John Young on the Moon in 1972

If you visited other planets, or the Moon, your mass (the amount of material you are made of) would stay the same. But your weight is determined by the pull of gravity on mass, and each planet has a different gravitational pull. The greater the pull, the greater your weight. To find your weight on another planet or the Moon, multiply your Earth weight by the percentage shown in the table above, and divide the answer by 100.

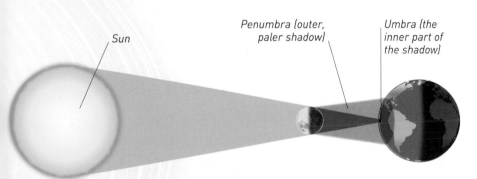

Sun

Penumbra (outer, paler shadow)

Umbra (the inner part of the shadow)

What is a solar eclipse?
The Sun is 400 times larger than the Moon, but 400 times more distant, so they appear the same size in our sky. When the Sun, Moon, and Earth are directly aligned (above), the Sun is eclipsed, and hidden from view. The shadow cast by the Moon falls on Earth. In the darkest part – the umbra – darkness falls and a total solar eclipse is seen. Anyone in the penumbra – the rest of the shadow – sees a partial eclipse.

VIEW A SOLAR ECLIPSE

Date	Visible from this location
21/08/17	North Pacific, USA, South Atlantic
02/07/19	South Pacific, Chile, Argentina
14/12/20	South Pacific, Chile, Argentina, South Atlantic
04/12/21	Antarctica
08/04/24	Mexico, central USA, east Canada
12/08/26	Arctic, Greenland, Iceland, Spain
02/08/27	Spain, north Africa, Saudi Arabia, Yemen

USEFUL WEBSITES

- For all types of NASA space missions – past, present, and future – look at *http://www.jpl.nasa.gov/missions/*
- Follow Curiosity as it roves across Mars, on *http://mars.nasa.gov/msl/* and receive its tweets on *https://twitter.com/MarsCuriosity*
- Find out how the Kepler spacecraft searches for exoplanets, and learn about its latest discoveries, on *http://kepler.nasa.gov/*
- Play games and learn more about space on *www.nasa.gov/kidsclub/index.html* and *http://www.esa.int/esaKIDSen/*

PLACES TO VISIT

VISIT A SPACE CENTRE
- At the Kennedy Space Center, Florida, USA, see the rockets that launched astronauts into space, and much more. *https://www.kennedyspacecenter.com/*

VISIT A MUSEUM
- The National Space Centre, Leicester, UK, is packed with interactive galleries, a Rocket Tower, and a planetarium. *http://spacecentre.co.uk/*
- The Herschel Museum of Astronomy, Bath, UK, was William Herschel's home when he discovered Uranus in 1891. *http://herschelmuseum.org.uk/*
- The Cité de l'espace, Toulouse, France, is a theme park dedicated to space. *http://en.cite-espace.com/discover-the-cite-de-lespace/*

VISIT AN OBSERVATORY
- Palomar Observatory, California, USA, is home to the 200-in (5-m) Hale telescope. *http://www.astro.caltech.edu/palomar/visitor/*

VISIT AN IMPACT CRATER
- Walk around the rim and peer deep inside the Barringer Meteorite Crater, Arizona, USA. *http://www.barringercrater.com/*
- See Earth's second-largest impact crater at Wolfe Creek, Western Australia. *https://parks.dpaw.wa.gov.au/park/wolfe-creek-crater*

Uranus

19 au

Neptune

30 au

Timeline

Earth's sky is our window on the Solar System. Our earliest ancestors followed the motions of the Sun, Moon, and planets. Later, with telescopes, people saw the details on these objects, and discovered many more. Spacecraft then became our eyes and laboratories in space, opening up these far-off worlds.

4000 BCE   **500** BCE

Nebra sky disc, Germany, c.1600 BCE

550 BCE
Greek mathematician Pythagoras says the Sun, Moon, Earth, and planets are spherical, not flat.

Pythagoras

4000 BCE
Ancient cultures in the Near East use the Sun and Moon for time keeping, and think five planets – Mercury, Venus, Mars, Jupiter, and Saturn – orbit Earth.

1900

Yuri Gagarin, inside Vostok 1

1957
The first spacecraft, Sputnik 1, is put into orbit around Earth by the Soviet Union.

Model of Sputnik 1

1930
American astronomer Clyde Tombaugh discovers a ninth planet – Pluto. Its status is changed to that of dwarf planet in 2006.

1801
Ceres is the first asteroid to be found, by Italy's Giuseppe Piazzi. It is also a dwarf planet.

1961
Soviet cosmonaut Yuri Gagarin becomes the first human in space, orbiting Earth in Vostok 1.

1959
The Soviet Union's Luna 2 is the first to land on the Moon.

1846
German astronomer Johann Galle discovers Neptune, very close to the position predicted by French mathematician Urbain Le Verrier.

Ceres

1960   **1980**

1962
American spacecraft Mariner 2 makes the first successful flyby of a planet, Venus. Mariner 4 is the first to fly by Mars, in 1965.

1971
Mariner 9 is the first craft to orbit a planet other than Earth. It orbits Mars for almost one year.

1979
The first flyby of Saturn is made by Pioneer 11.

1986
American craft Voyager 2 becomes the first and only craft to fly by Uranus. In 1989, it becomes the only spacecraft to fly by Neptune.

1973
American craft Pioneer 10 is the first to fly by Jupiter. In 1974, Mariner 10 makes the first flyby of Mercury.

1976
The first craft to land on Mars is America's Viking 1, which looks for signs of life.

Neil Armstrong on the Moon, 1969

1969
On 21 July, American astronaut Neil Armstrong steps on to the Moon, becoming the first human to walk on another world.

1970
The first rover to land on another world, the Soviet Union's Lunokhod 1 explores the Moon for 10 months.

1975
The Soviet Union's Venera 9 transmits the first images from the surface of Venus.

Pioneer 11 image of Saturn

1252
The Alphonsine Tables produced by scholars for Alfonso X of Castile list the accurate positions of the Sun, Moon, and planets for specific dates.

964 CE
Persian scholar Abd al-Rahman al-Sufi updates the astronomy of Ancient Greece in his *Book of the Fixed Stars*.

1424
Islamic astronomer Ulugh Beg builds his observatory at Samarkand (now in Uzbekistan).

Persian depiction of the Perseus constellation, from the *Book of the Fixed Stars*

Nicolaus Copernicus

1596
The last great naked-eye astronomer, Danish Tycho Brahe completes 20 years of observations. German mathematician Johannes Kepler uses them to form his three laws of planetary motion in 1609–19.

1543
Polish astronomer Nicolaus Copernicus suggests the Sun is at the centre of the Universe, orbited by Earth and the other planets.

Replica of Herschel's telescope

1781
German-born English astronomer William Herschel discovers a new planet – Uranus – doubling the known extent of the Solar System.

1687
English scientist Isaac Newton's theory of gravity is published. It explains why the Moon orbits Earth and the planets orbit the Sun.

1682
English scientist Edmond Halley sees a comet, calculates its orbit, and predicts its return. Named after him, it returns every 76 years.

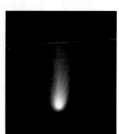

Halley's Comet, 1986

1655
Dutch scientist Christiaan Huygens observes Saturn and says it is encircled by a ring. In 1659, he measures the spin period of Mars.

1609
The newly invented telescope is used by Italian astronomer Galileo Galilei to study the Moon, planets, and stars. In 1610, he publishes his discoveries.

Galileo Galilei

1992
The first Kuiper Belt object – 1992 QB1 – is discovered by English astronomer David Jewitt and Vietnamese American astronomer Jane Luu.

51 Pegasi b

1995
The first exoplanet orbiting a Sun-like star is detected – 51 Pegasi b.

1991
Gaspra is the first asteroid encountered by a spacecraft when the American mission Galileo flies by en route to Jupiter.

2006
Pluto and Eris are reclassified as dwarf planets.

2012
The American rover Curiosity lands on Mars, and studies whether or not Mars could have supported life in the past.

2014
The European Space Agency's Rosetta orbits a comet, and releases Philae, the first craft to land on a comet nucleus.

2016
Juno, an American spacecraft, becomes the second craft to orbit Jupiter, and returns its first images from the planet.

2015
American craft New Horizons flies by Pluto before it heads to Kuiper Belt object 2014 MU69.

Juno spacecraft

Glossary

ANTENNA
An aerial in the shape of a rod, dish, or array for receiving or sending radio signals.

ASTEROID
A small rocky body orbiting the Sun. Most asteroids are in the Asteroid Belt, between Mars and Jupiter.

ASTEROID BELT
A region in the Solar System, between the orbits of Mars and Jupiter, that contains a large number of orbiting asteroids.

ASTRONOMICAL UNIT (AU)
A unit of distance of 149,597,870,700 m – about 149.6 million km (92.9 million miles) – Earth's average distance from the Sun.

ATMOSPHERE
The layer of gases held around a planet, moon, or star by its gravity.

AURORA
A colourful light display above a planet's polar regions – such as Earth's Northern and Southern Lights – produced when electrically charged particles hit atoms in the planet's atmosphere and make it glow.

Northern Lights (Aurora Borealis) over a lagoon in Iceland

AXIS
The imaginary line that passes through the centre of a space object, such as a planet or star, and around which the object rotates, or spins.

BILLION
One thousand million (1 followed by nine zeros).

COMET
A small body made of snow, ice, and rock dust (called the nucleus). Comets that get deflected from their orbits in the outer Solar System and travel near the Sun develop a large cloud of gas and dust (the coma) and gas and dust tails.

CORE
The innermost region of a planet or star.

CRATER
A dish-shaped hollow on the surface of a rocky planet, moon, or asteroid. An impact crater is made by an asteroid, meteorite, or comet hitting the surface. A volcanic crater develops as a volcano ejects material.

CRUST
The thin, outermost layer of a rocky or icy body, such as a planet, moon, or comet.

DWARF PLANET
An almost round body that orbits the Sun but does not have enough mass and gravity to clear its neighbourhood of other objects.

ECLIPSE
The effect observed when two space bodies are aligned so one appears directly behind the other, or one is in the shadow of the other. In a solar eclipse, when the Moon passes between Earth and the Sun, the Sun is behind the Moon in the sky. In a lunar eclipse, the Moon passes behind Earth, through Earth's shadow.

EQUATOR
The imaginary line drawn around the middle of a planet, moon, or star, halfway between its north and south poles, separating its northern and southern hemispheres.

EXOPLANET
A planet that orbits around a star other than the Sun.

FLYBY
A close encounter made with a Solar System object by a spacecraft that flies past without going into orbit.

GALAXY
An enormous grouping of stars, gas, and dust held together by gravity. The Sun is one of the stars in the Milky Way Galaxy.

GAS GIANT
A large planet, such as Jupiter or Saturn, that consists mainly of hydrogen and helium, which are in gaseous form at the planet's visible surface.

GIANT PLANET
Any planet that is large and massive compared with Earth. In the Solar System, the four biggest planets – Jupiter, Saturn, Uranus, and Neptune – are classed as giant planets.

GRAVITY
A force of attraction found throughout the Universe. The greater the mass of a body, the greater its gravitational pull.

HABITABLE
Suitable for living in or on. Life may exist on a planet that orbits within the habitable zone around a star.

The doomed hot Jupiter WASP-12b is being eaten by its parent star

HEMISPHERE
One half of a sphere. Earth is divided into northern and southern hemispheres by its equator.

HOT JUPITER
A type of exoplanet that is more massive than Jupiter but orbits closer to its star than Mercury to the Sun.

ICE GIANT
A giant planet composed mainly of elements heavier than hydrogen and helium. The Solar System's two ice giants are Uranus and Neptune.

INNER PLANETS
The planets nearest the Sun in the Solar System – Mercury, Venus, Earth, and Mars.

INTERSTELLAR SPACE
The region between the stars in a galaxy – in the Solar System, it is the space beyond the planets.

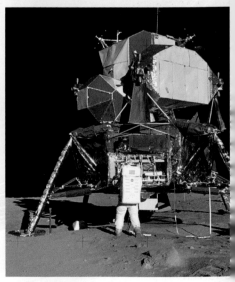

Apollo 11's lunar module, Eagle – the first lander to put people on the Moon

KUIPER BELT OBJECT (KBO)
A rock-and-ice body orbiting the Sun in the Kuiper Belt, beyond Neptune's orbit.

LANDER
A spacecraft that lands on the surface of a planet, moon, asteroid, or comet.

LUNAR
Relating to the Moon, such as a lunar module built to land on the Moon.

MAGNETIC FIELD
The region around a magnetized body, where magnetic forces affect the motion of electrically charged particles. Earth's magnetic field, for example, is generated by flows in the planet's liquid outer core.

MANTLE
A thick layer of rock between the core and the crust of a planet or moon.

MASS
A measure of how much matter (material) a body is made of.

METEOR
A short-lived streak of light – also called a shooting star – produced by a tiny, grain-sized piece of comet or asteroid speeding through Earth's upper atmosphere.

METEORITE
A piece of rock or metal that lands on the surface of a planet or moon and survives the impact; most are pieces of asteroid.

Incoming meteorite

MILKY WAY
The spiral-shaped galaxy that includes the Sun and about 200 billion other stars. Also the stars visible to the naked eye as a band of faint light across the night sky.

MOON
A rock or rock-and-ice body that orbits a planet, dwarf planet, or asteroid.

NUCLEAR FUSION
A process in which atomic nuclei join to form heavier nuclei and release huge amounts of energy such as heat and light. In the Sun's core, hydrogen nuclei fuse to produce helium.

NUCLEUS (plural NUCLEI)
The compact central core of an atom. Also the solid, icy body of a comet.

OORT CLOUD
The sphere consisting of billions of comets that surrounds the planetary part of the Solar System.

ORBIT
The path of a natural or artificial object around another more massive body, influenced by its gravity.

ORBITER
A spacecraft that orbits a space body such as a planet, moon, or asteroid.

OUTER PLANETS
The four planets that orbit the Sun beyond the Asteroid Belt – Jupiter, Saturn, Uranus, and Neptune.

PARTICLE
An extremely small part of a solid, liquid, or gas.

PHASE
The part of a moon or planet that is lit by the Sun and visible from Earth. The Moon passes through a cycle of phases every 29.5 days.

PLANE
A flat, two-dimensional area. In the Solar System, the planets travel around the Sun at different distances but close to the same orbital plane.

PLANET
A massive, nearly round body that orbits the Sun and shines by reflecting the star's light.

PROBE
An unmanned spacecraft built to explore objects in space – particularly their atmosphere and surface – and transmit information back to Earth.

ROCKY PLANET
A planet composed mainly of rock, such as the four planets closest to the Sun – Mercury, Venus, Earth, and Mars.

ROVER
A vehicle driven remotely on a planet or moon.

SATELLITE
An artificial object that is deliberately placed in orbit around Earth or another Solar System body. Also a natural object such as a moon or any space object orbiting another, more massive body.

SHEPHERD MOON
A small moon whose gravitational pull herds orbiting particles into a well-defined ring around a planet.

SOFT LANDING
A controlled landing by a vehicle, including a spacecraft. A hard landing may result in the craft being destroyed, intentionally or accidentally.

SOLAR
Relating to the Sun.

SOLAR SYSTEM
The Sun and all the objects orbiting it, such as the planets and many smaller bodies.

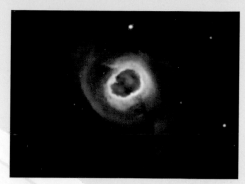

Dying star, a glimpse of our Sun's future

STAR
A huge, spinning sphere of very hot, luminous gas that generates energy by nuclear fusion in its core. The Sun is a star.

SUPER-EARTH
A type of exoplanet with a mass greater than Earth but smaller than the ice giants Uranus and Neptune.

TELESCOPE
An instrument that uses lenses or mirrors, or a combination of the two, to collect and focus light to form a magnified image of a distant object. Some telescopes collect other wavelengths besides visible light, such as radio and infrared.

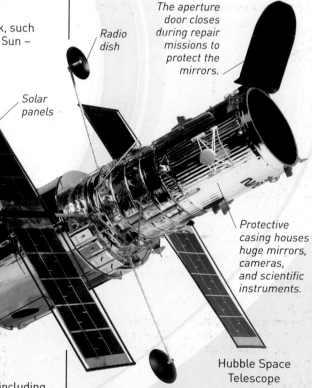

The aperture door closes during repair missions to protect the mirrors.

Radio dish

Solar panels

Protective casing houses huge mirrors, cameras, and scientific instruments.

Hubble Space Telescope

UNIVERSE
Everything that exists; all space and everything in it.

VOLCANISM
The eruption of molten rock on to the surface of a planet or moon through volcanic vents, often resulting in extensive flows of lava.

Index

Acknowledgements

Dorling Kindersley would like to thank:
Carron Brown for the index, Victoria Pyke for proofreading, Antara Raghavan for editorial assistance, and Mahua Mandal and Sanjay Chauhan for design assistance.

The publisher would like to thank the following for their kind permission to reproduce their photographs:

(Key: a-above/bottom; b-below/bottom; c-centre; f-far; l-left; r-right; t-top)

2 NASA: JPL-Caltech / Hap McSween (University of Tennessee), and Andrew Beck and Tim McCoy (Smithsonian Institution) (bl). Science Photo Library: Take 27 Ltd (t). 3 NASA: JHUAPL / SwRI (c). 4 NASA: (l); JPL (c); JPL-Caltech / SSI / Hampton University (crb). 4-5 Dorling Kindersley: Simon Mumford (t). 5 NASA: JPL-Caltech / University of Arizona (tr). 6 ESA: NASA / JHU Applied Physics Lab / Carnegie Inst. Washington (c). NASA and The Hubble Heritage Team (AURA/STScI): NASA / ESA / A. Simon (Goddard Space Flight Center) (bl). NASA: (crb, c); JPL-Caltech (fcrb). 7 Eliot Herman: (tr). NASA: JPL / Space Science Institute (clb). 8 NASA: Johns Hopkins University Applied Physics Laboratory / Southwest Research Institute (cl). 9 Dreamstime.com: Typhoonski (cl). NASA: Cosmos Studios (c); Johns Hopkins University Applied Physics Laboratory / Carnegie Institution of Washington (t). 10 123RF.com: ammit (b). NASA and The Hubble Heritage Team (AURA/STScI): ESA / A. Simon (Goddard Space Flight Center) (fcrb); ESA / R. Beebe (New Mexico State University) (cb); ESA / H. Hammel (Space Science Institute and AURA) (crb). NASA: JPL-Caltech / Cornell (ASU) (clb). 10-11 Science Photo Library: Take 27 Ltd (c). 11 ESA: J. Whatmore (tr). NASA: (crb); Johns Hopkins University Applied Physics Laboratory / Carnegie Institution of Washington (c). 12 Alamy Stock Photo: Granger Historical Picture Archive (tc); North Wind Picture Archives (t). 12-13 ESO: Inigocia (b). 13 ESO: Y. Beletsky/http://creativecommons.org/licenses/by/3.0 (cr). NASA and The Hubble Heritage Team (AURA/STScI): ESA / H. Teplitz and M. Rafelski (IPAC / Caltech) /

A. Koekemoer (STScI) / R. Windhorst (Arizona State University) (cla). Library of Congress, Washington, D.C.: Whipple, John Adams (cb). 14 NASA: (cra); JPL-Calech / University of Arizona (cr). 15 NASA: (cla, l); JPL (br); JPL-Caltech / MSSS (tr); Canberra Deep Space Communication Complex (cb). 16 Dreamstime.com: Oleg Znamenskiy (crb). NASA: SDO (l). 17 Dreamstime.com: Igor Kovalchuk (bl). Getty Images: Heritage Images / Hulton Archive (t). NASA: ESA (tr); ESA / SOHO (cr); SDO (cl). Science Photo Library: Detlev Van Ravensswaay (crb). SOHO (ESA & NASA): Alex Lutkus (tc). 18-19 NASA: JHU Applied Physics Lab / Carnegie Inst. Washington (c). 18 ESO: Y. Beletsky/http://creativecommons.org/licenses/by/3.0 (cl). 19 ESA: (br). NASA: (tc, crb); Aubrey Gemignani (tc/MESSENGER); Johns Hopkins University Applied Physics Laboratory / Carnegie Institution of Washington (cra). 20-21 JPL (c). 20 NASA: (cla, bl). 21 NASA (crb). 22-23 NASA: (bc). 23 Getty Images: Arlan Naeg / AFP (crb). NASA: (bc). 24 NASA: (cl). 24-25 Dorling Kindersley: Simon Mumford (c). 25 Getty Images: Universal Images Group (t). NASA: JPL (cr). 26 FLPA: Frans Lanting (b). Getty Images: Gail Shumway (t). NASA: (cl). 27 Dreamstime.com: Sean Beckett (br); Heisenberg85 (t). NASA: Jesse Allen / NOAA / UNEP (cl). Science Photo Library: Dr Seth Shostak (clb). 28 NASA: JPL-Caltech (cl). 29 Dreamstime.com: Delstudio (c); Sonsam (bc). ESO: D. Schreiner and S. Degezelle/http://creativecommons.org/licenses/by/3.0 (cla). 30 Getty Images: Stocktrek Images (r). NASA: (cl); GSFC / Arizona State University (cra). 30-31 NASA: (b). 31 NASA: (cr, t); Ulli Lotzmann (cla). Wikipedia: (cl). 32-33 NASA: JPL (c). 32 ESA: DLR / FU Berlin (G. Neukum) (br). NASA: (t). 33 ESA: DLR / FU Berlin, CC BY-SA 3.0 IGO (bl); R. Lockwood (br). NASA: JPL-Caltech / University of Arizona (cla, cra, cl). 34 Dreamstime.com: Pixattitude (tl). 34-35 NASA: JPL (tc). Science Photo Library: JPL / ARIZONA STATE UNIVERSITY (b). 35 ESA: DLR / FU Berlin (G. Neukum) (tr). NASA: JPL / Cornell (c). 36 NASA: JPL-Caltech (tr). 36-37 NASA: JPL-Caltech / MSSS (b). 37 ESA: ATG medialab (cr). NASA: JPL-Caltech / LANL /

J.-L. Lacour, CEA (tl); JPL (tr); JPL-Caltech / MSSS (c). 38 NASA: JPL-Caltech / UCLA / MPS / DLR / IDA (t); JPL-Caltech / Hap McSween (University of Tennessee), and Andrew Beck and Tim McCoy (Smithsonian Institution) (cr). 39 Alamy Stock Photo: dpa picture alliance archive (bc). ESO: NASA/Jeff Schmaltz/http://creativecommons.org/licenses/by/3.0 (cr). Getty Images: Thomas J. Abercrombie (bl). NASA: M. Ahmetvaleev (t). NASA and The Hubble Heritage Team (AURA/STScI): NASA / ESA / A. Simon (Goddard Space Flight Center) (ca); NASA / ESA / J. Nichols (University of Leicester) (b). NASA: NSSDC / GSFC (clb). 41 NASA: JPL / Cornell University (clb); JPL (cl); JPL / Space Science Institute (tr). 42 NASA: JPL / DLR (cr). 42-43 NASA: JPL-Caltech / KSC (tr). 43 NASA and The Hubble Heritage Team (AURA/STScI): ESA / J. University of Arizona (cla). Copyright © Subaru Telescope, National Astronomical Observatory of Japan (NAOJ). All right reserved.: (b). 44 Getty Images: Print Collector / Hulton Archive (tt). 45 NASA: JPL-Caltech / Space Science Institute (tr, bc); JPL / Space Science Institute (tc); JPL-Caltech / SSI / Hampton University (crb). 46 Getty Images: De Agostini Picture Library (br). NASA: JPL / Space Science Institute (cra). 47 NASA: JPL-Caltech / Keck (clb); JPL / Space Science Institute (tr, cra); The Hubble Heritage Team (STScI / AURA)Acknowledgement: R.G. French (Wellesley College), J. Cuzzi (NASA / Ames), L. Dones (SwRI), and J. Lissauer (NASA / Ames) (tc). 48 NASA: JPL-Caltech / Space Science Institute (tr); JPL / Space Science Institute (cr). 49 NASA: (cl); ASI / Cornell (tl); ESA (cl). 50 NASA: (cl); Don Davis (br); GRIN (cl). 51 NASA: JPL (ca); JPL-Caltech (crb). 52 ESO: http://creativecommons.org/licenses/by/3.0 (bc). Getty Images: John Russell (bl). 53 Getty Images: Jonathan Blair (crb). NASA and The Hubble Heritage Team (AURA/STScI): ESA / Adolf Schaller (bc). NASA: R. Hurt (SSC-Caltech) / JPL-Caltech (clb); JHUAPL / SwRI (cb). 54 NASA and The Hubble Heritage Team (AURA/STScI): Erich Karkoschka (University of Arizona) (clb). 55 ESO: http://creativecommons.org/licenses/by/3.0 (b). NASA and The Hubble Heritage Team (AURA/STScI): ESA / M. Showalter (SETI Institute) (tr). NASA: JPL (tc, bc). 56 NASA and The Hubble Heritage Team (AURA/STScI): ESA / M.H. Wong / J. Tollefson (UC Berkeley) (cla, cl). 57 NASA: JPL (cl, bc). 58 NASA: USGS (tr). Dr Dominic Fortes, UCL: (cra). 58 Getty Images: Bettmann (cb). NASA: JHUAPL / SwRI (b). 58-59 NASA: JHUAPL / SwRI (c). 59 ESO: L. Calçada and Nick Risinger/http://creativecommons.org/licenses/by/3.0 (cb/Eris).

NASA and The Hubble Heritage Team (AURA/STScI): ESA / Adolph Schaller (for STScI) (br). NASA: (clb/Haumea); Johns Hopkins University Applied Physics Laboratory / Southwest Research Institute (cb); JHUAPL / SwRI (clb/Pluto). 60 ESO: http://creativecommons.org/licenses/by/3.0 (bl). Getty Images: Dea / A. Dagli Orti (bc). 61 ESA: ATG medialab (cb). NASA: (ca); ESA / Rosetta / NAVCAM (b). NASA and The Hubble Heritage Team (AURA/STScI): ESA / P. Kalas (University of California, Berkeley) (ti). NASA: JPL-Caltech (br, cl). 63 NASA: ESA / A. Feild (STScI) (bc); Ames / SETI Institute / JPL-Caltech (c); JPL-Caltech / Palomar Observatory (clb). 64 Getty Images: Archive Photos (cra). NASA: (cla); JPL / Cornell University (b); JPL (c). 64-65 NASA: JPL (Background). 65 Dreamstime.com: Inge Hogenbijl (cr). NASA and The Hubble Heritage Team (AURA/STScI): ESA / A. Simon-Miller (NASA Goddard Space Flight Center) (ca). NASA: (clb). 66-67 NASA: JPL (Background). 67 NASA: (ca). 68 Getty Images: Heritage Images / Hulton Fine Art Collection (tc). NASA: (cla, br, clb); JPL-Caltech / UCLA / MPS / DLR / IDA (cr). Wellcome Images http://creativecommons.org/licenses/by/4.0/: Iconographic Collections (tc). 68-69 NASA: JPL (Background). 69 Dreamstime.com: Andrey Andronov (ca). ESO: http://creativecommons.org/licenses/by/3.0 (c); M. Kornmesser/Nick Risinger (skysurvey.org)/http://creativecommons.org/licenses/by/3.0 (b). Getty Images: Dea Picture Library (cr); Heritage Images / Hulton Fine Art Collection (tl); Science & Society Picture Library (cla). NASA: JPL-Caltech (b). 70 Dreamstime.com: Jamen Percy (clb). NASA: Neil Armstrong (br); ESA / G. Bacon (tr). 70-71 NASA: JPL (Background). 70-71 Dorling Kindersley: Andy Crawford (crb). ESA: NASA / The Hubble Heritage Team (STScI / AURA) / R. Sahai and J. Trauger (Jet Propulsion Laboratory) (tr)

Wallchart: ESA: JHU Applied Physics Lab / Carnegie Inst. Washington tr; NASA and The Hubble Heritage Team (AURA/STScI): cr; NASA: b, JPL-Caltech / MSSS ca, Johns Hopkins University Applied Physics Laboratory / Southwest Research Institute cb, Johns Hopkins University Applied Physics Laboratory / Southwest Research Institute cb/ (Pluto), JPL-Caltech crb, JPL-Caltech cra, SDO tl

All other images © Dorling Kindersley
For further information see: www.dkimages.com